With compliments

INTERNATIONAL REVIEWS IN

AEROSOL PHYSICS AND CHEMISTRY

VOLUME 2

TOPICS IN CURRENT AEROSOL RESEARCH

TOPICS IN CURRENT AEROSOL RESEARCH

EDITED BY

G. M. HIDY

*Science Center, North American Rockwell Corporation,
Thousand Oaks, California 91360*

AND

J. R. BROCK

University of Texas, Austin, Texas

PERGAMON PRESS

Oxford · New York · Toronto

Sydney · Braunschweig

Pergamon Press Ltd., Headington Hill Hall, Oxford
Pergamon Press Inc., Maxwell House, Fairview Park, Elmsford, New York 10523
Pergamon of Canada Ltd., 207 Queen's Quay West, Toronto 1
Pergamon Press (Aust.) Pty. Ltd., 19a Boundary Street, Rushcutters Bay, N.S.W. 2011, Australia
Vieweg & Sohn GmbH, Burgplatz 1, Braunschweig

First edition 1971
Library of Congress Catalog Card No. 70–104120

PRINTED IN GREAT BRITAIN BY THE WHITEFRIARS PRESS LTD, LONDON AND TONBRIDGE
08 016674 1

CONTENTS

PREFACE

DESPITE an ever-widening technological interest in aerocolloidal systems, the fundamental science of aerosols, on which practical considerations must rely, largely has been set aside until recently in deference to other problems. However, in the last decade a kind of renaissance in aerosol research has taken place in which much of the classical work of the early twentieth century is being extended. New sophisticated theoretical and experimental techniques are being developed, and are being applied to understand better the behavior of aerosol systems.

As scientists in many diverse fields expand their activity in aerosol research, the results of investigations appear in a wide variety of journals that reach entirely different small groups of workers. The problems of communication of these scientific results are complicated further by the worldwide character of aerosol science. To help focus attention on the variety of important aerosol research presently being published, and to open a new channel for international communications between workers in this field, we have organized a new series entitled *International Reviews in Aerosol Physics and Chemistry*. This work will consist of a collection of monographs of book length, and companion volumes of selected review articles dealing with several aspects of aerosol science, and its relationship to the study of the so-called "particulate state of matter." The scope of the *International Reviews* will be limited to results which contribute significantly to the state of fundamental knowledge of aerosol behavior.

Because *International Reviews in Aerosol Physics and Chemistry* is designed to concentrate on the fundamental aspects of aerosol science, it should have considerable usefulness both to practising scientists and to graduate students in such widely diverse fields as physics, physical chemistry, meteorology, geophysics, astronomy, chemical engineering, mechanical engineering, aerospace engineering, environmental sciences, and medicine.

As the second volume of the series, this book includes two separate articles on aerosol science, written by well-known investigators. The material presented in these two reviews represents a viewpoint on different aspects of several problems. Prof. Fuchs and Dr. Sutugin discuss experiment and theory describing highly dispersed aerosols. Prof. Soo's article gives a capsule review of his and his colleagues' work devoted to expounding on a "transport" model for electrically charged aerosols.

In their article, Fuchs and Sutugin describe several properties of very small aerosol particles, the means of generation of such suspensions, and the dynamical behavior of such particles, either as single, idealized spheres, or in collections. The article gives somewhat broader coverage to the experimental and theoretical efforts in this subject than in the first volume of this series. The significance of non-continuum dynamical effects is emphasized, as well as the peculiarities of experimental methods required in this subject.

Prof. Soo summarizes in his contribution much of his efforts over the past several years to construct an aerosol model based on an extension of the theory of transport in multicomponent gas systems. In his model, aerosol particles of a given size class are considered a component in the gas mixture in analogy to a gas molecule. Prof. Soo illustrates his views with a variety of experimental work, on both uncharged and electrically charged particle species in gases. It is particularly interesting to note in passing the "success" of boundary layer treatments of some experimental observations of flowing aerosol suspensions. Similar results were discussed in a different way in Chapter 7 of Volume 1.

We are very grateful to the authors for their participation in the *International Reviews*. It is a great pleasure to acknowledge their extensive efforts and patience in preparation of the manuscripts for publication.

1970

G. M. Hidy
J. R. Brock

HIGH-DISPERSED AEROSOLS

N. A. FUCHS and A. G. SUTUGIN

Karpov Institute of Physical Chemistry, Moscow

Abstract. The processes leading to the formation of high-dispersed aerosols with particle size below 0.1 μ and the methods of generation and investigation of these aerosols are treated as well as their physical properties differing fundamentally from those of coarse aerosols. In spite of the great role played by high-dispersed aerosols in the important process of vapor condensation in space and their significance for many industries and meteorology, there are no monographs or reviews on this subject available in the literature, and the aim of the present work is to fill this gap.

CONTENTS

INTRODUCTION

BASIC PROPERTIES OF HIGH-DISPERSED AEROSOLS (HDAs)

IN THIS review high-dispersed aerosols (HDAs) are understood to refer to aerodisperse systems with particle size† less than 1000 Å. At atmospheric pressure in air this corresponds to Knudsen number $Kn > 1$ (Kn is the ratio of the mean free path of gas molecules to particle radius). The most important property of HDAs with $Kn \gg 1$ is that the momentum, energy, and mass transfer processes from particles to medium and vice versa are described by gas-kinetic formulas, i.e. particles can be considered as gigantic gas molecules. Therefore the resistance of the medium to the motion of particles, the rates of their evaporation (mass decrease), and of heat transfer to them, as well as the thermophoresis rate, are proportional to the particle radius squared. The aerosol Knudsen number, i.e. the ratio of the apparent mean free path of the particles themselves to their radius at $Kn \gg 1$, is also greater than unity and, hence, the coagulation rate of HDAs is expressed by the formula for the number of collisions between gas molecules, and the coagulation constant increases with particle size.

For HDAs with Kn not much greater than unity, some correction factors, increasing with decreasing Kn, are to be introduced into the above-mentioned formulas. At small Kn, however, i.e. in coarse-dispersed aerosols, the transfer processes are governed by quite different laws based on the equations of hydrodynamics, heat conductivity, and diffusion in continuous media.

The vapor pressure of HDA particles is appreciably greater than the equilibrium pressure over the plane surface of a particle substance. The light scattering by HDAs obeys the Rayleigh equation, the intensity of scattered light being so low that it is practically impossible to examine HDAs by macro-optical methods, whereas under normal illumination and observation conditions HDA particles are unobservable in an ultra-microscope.

The Brownian motion of HDA particles is so vigorous and their inertia and sedimentation rate due to gravity are so small, that HDA deposition on walls or obstacles, etc., occurs exclusively by diffusion. The charges acquired by the particles of $r < 0.03\ \mu$ in a bipolarly ionized atmosphere do not exceed one elementary charge, the proportion of charged particles decreasing with their size.

In all the properties discussed above, HDAs differ greatly from aerosols of particle size greater than 1000 Å, which so far have been the main subject of studies. It should be also noted that according to what has been said above about the charges of HDA particles, their electric mobility is greater than 10^{-4} cm^2/v sec. Thus, to use the terms of aerophysics, charged atmospheric HDAs belong to "medium" and "heavy" atmospheric ions.

† Particle "size" d is understood to mean the diameter for spherical particles or an average diameter for non-spherical ones. In some cases, however, particle radius a is given.

CHAPTER 1

FORMATION AND METHODS OF GENERATION OF HDAs

1.1. Formation of condensation HDAs

In the treatment of the theory of formation of HDAs it should be considered first under what conditions vapor condensation results in the formation of very fine particles. The mean particle mass is equal to the ratio of the condensed vapor mass to the number of particles formed. It follows from any one theory of spontaneous vapor condensation that the rate of formation of new phase nuclei rises faster than the first power of supersaturation, whereas the rate of condensation growth of particles is approximately proportional to supersaturation. Therefore the mean particle size in the aerosol formed, in the absence of coagulation, should be the smaller the higher the supersaturation, i.e. the greater the supercooling of vapor (in the case of an instantaneous temperature drop). In reality, cooling occurs at a certain finite rate, and aerosol formation, including nucleation and condensation growth of nuclei, occurs throughout the whole cooling time. The faster the vapor cools at the same initial and final temperatures, the lesser proportion of vapor is consumed in the growth of particles formed at earlier cooling stages at relatively low supersaturations, and the greater is the number of particles formed.

As regards the influence of the particle substance nature on the aerosol particle size, of most importance is the steepness of the temperature logarithm of vapor pressure curve. Under the same conditions of vapor cooling, substances with a steeper curve give more highly dispersed aerosols. For a given substance, the steepness of the curve increases with decreasing temperature. Therefore if the vapor saturated at temperature T_1 and T_2 ($T_1 > T_2$) is instantaneously cooled by ΔT degrees, in the second case a higher supersaturation and a more highly dispersed aerosol will occur. Here, in addition to the high supersaturation attained, the decrease of the equilibrium vapor concentration in the aerosol formed is important. If this concentration is not too low, HDAs are extremely unstable, and the process of larger particles "devouring" smaller ones is rapid. This process is due both to isothermal distillation of vapor from small to large particles and to the effect of local temperature fluctuations in aerosol. Already, with a very slight local temperature rise, smaller particles evaporate completely, the liberated vapor condensing on larger particles upon continued cooling. Therefore, in the case of substances volatile at room temperature, such as water, HDAs can be obtained only by rapid cooling to the liquid nitrogen temperature.

Generally, formation of HDA occurs by spontaneous condensation since the concentration of foreign condensation nuclei (gas ions included) is small compared to that of the nuclei formed spontaneously at high supersaturation. However, in the case of HDA formation in plasma burners, during cesium vapor condensation in a corona discharge in the case of flow through a Laval nozzle, etc. (i.e. at very high concentrations of gas ions), condensation seems to occur on the ions.

In the foregoing the particles' coagulation involving their coalescence or sintering was not considered. However, as will be shown below, coagulation plays a fundamental role in HDA formation at medium vapor concentrations.

The classical theory of nucleation[1] has been extensively covered in the literature. Therefore only the statement of the basic assumptions of the theory will be considered. The supersaturated vapor is assumed to consist of a mixture of single molecules and molecular clusters, the concentration of which, according to the general theory of thermodynamic fluctuations, is an exponential function of the free enthalpy of their formation ΔG. The dependence of ΔG on the number of molecules in the cluster g in supersaturated vapor has a maximum. To a certain extent, each cluster can increase in size by adding molecules colliding with it or decrease by evaporating molecules. For clusters of the size greater than the critical value g^* corresponding to the maximum ΔG, the probability of growth is much larger than that of evaporation. Such supercritical clusters are considered as being new phase nuclei. The change with time of the concentration of clusters containing g molecules is expressed by the equation

$$\Delta f_g/\Delta t = \alpha_{g-1} f_{g-1} - \alpha_g f_g + \beta_{g+1} f_{g+1} - \beta_g f_g, \tag{1.1}$$

where α_g is the probability of collision of a cluster with a vapor molecule and β_g the probability of evaporation of a molecule from the cluster. The quantity β_g in the classical theory is expressed in terms of α_{g-1} by means of the detailed equilibrium principle. When g is large enough, the set of equations of the form (1.1) for g from 1 to g^* can be substituted by the differential equation

$$\frac{\partial f(g,t)}{\partial t} = \frac{\partial}{\partial g}\left\{\alpha_g \frac{\partial \Delta G(g,t)}{\partial g} + \alpha_g \frac{\Delta G(g,t)}{kT}\frac{\partial \Delta G(g,t)}{\partial g}\right\}, \tag{1.2}$$

which is actually an equation of the diffusion of clusters along the continuous axis g. Under steady-state conditions, the nucleation rate J can be determined assuming the supercritical particles to be removed from the system and substituted by an equal mass of vapor by the solution of the equation

$$-J = \alpha_g \frac{\partial \Delta G(g)}{\partial g} + \alpha_g \frac{\partial \Delta G(g)}{\partial g}\frac{\Delta G(g)}{kT}. \tag{1.3}$$

Owing to high supersaturation values necessary for the formation of HDAs, g^* is usually less than 10. Therefore the classical method of expressing the free enthalpy of nuclei, regarded as droplets of a liquid with its bulk properties, is apparently invalid. Moreover, at small g the discreteness of the axis f increases, and it cannot be considered any longer as being continuous. Very often the process leading to supersaturation, i.e. cooling of vapor or of a vapor–gas mixture, runs parallel to the formation of new phase nuclei and growth of particles being formed. Under these conditions, temperature, vapor concentration, and supersaturation change continuously in the course of the process, which makes the solution of equations of the form (1.2) impossible.

In addition to the classical method, there is another approach to the nucleation problem, which became known as the constant-number theory. In this theory the value of g^* is assumed to be small, not much more than 2 and almost independent of the nucleation conditions. The probability of evaporation of a nucleus containing g^*+1 molecules is disregarded and the concentrations of subcritical clusters are considered to be equal to equilibrium concentrations. In the classical theory, however, the difference between steady-state and equilibrium concentrations of small clusters is also small. The intuitive concepts which formed the basis of the theory of constant nucleus size were first advanced by La Mer[2] and Courtney.[3] From these concepts Christiansen,[4] by solving a set of four equations of the form (1.1), attempted to estimate the time necessary for the establishment of equilibrium concentrations of subcritical clusters after supersaturation in a monomolecular vapor is reached. These estimates, however, were only qualitative, since the value of free enthalpy of formation of small clusters was unknown. Reed[5] proposed a method of calculating this quantity. He expressed the chemical potential of molecular clusters in terms of their partition functions. He assumed that clusters of each size exist only in the most stable spatial configuration, that interactions in them occur only between pairs of nearest neighboring molecules according to the Lennard–Jones equation, and that molecules in the clusters preserve their rotational degrees of freedom. Reed calculated the chemical potentials of nitrogen clusters with $g \leqslant 8$. For larger clusters calculations are more cumbersome, and the error due to the existence of various spatial cluster configurations is greater. This error can be eliminated by using, instead of Reed's approach, the methods of calculating cluster integrals for all clusters. However, even the most recent methods of these calculations are much more bulky than Reed's method.

Sutugin and Fuchs[6] suggested a method of estimating nucleation at high supersaturations based on the combination of the constant-number theory with Reed's calculations. In the general case, the process of aerosol formation can be described by a system of non-linear differential equations of the form

$$\frac{df_g}{dt} = -\sum_{i=1}^{\infty} K_{ig} f_g f_i + \sum_{j+k=g} K_{jk} f_j f_k - \beta_g f_g + \beta_{g+1} f_{g+1}. \tag{1.4}$$

Usually, such a system cannot be solved even with the help of a high-speed electronic computer. However, HDA formation involves some factors simplifying this task. First, the process occurs so fast that by the time the rate of formation of new nuclei decreases owing to the depletion of the vapor phase, only small particles have been able to form. Therefore the total number of equations (1.4) is no more than a few tens or, at most, several hundreds. Secondly, the concentrations of subcritical clusters can be assumed to correspond to equilibrium concentrations for the temperature and concentration existing at the moment. Finally, the probability of evaporation of supercritical particles can be ignored. It is easy to see the method of calculation based on numerical integration of equations of the form (1.4) is applicable to nucleation in associated vapors as well.

This method was used to calculate the condensation of silver vapors contained in a hot turbulent argon jet issuing into cold stationary air in the range of silver vapor concentrations at the nozzle outlet 5×10^{14} to 5×10^{16} molecules/cm^2. The size of the critical nucleus was determined from the condition that the equilibrium concentration corresponding to this size and calculated by Reed's method was minimal.

Under the above conditions, the size of critical nuclei was 2–6 atoms, depending on temperature and concentration, so that the term "the constant-number theory" does not seem to be justified. The concentration of the particles formed being very large (10^{13}–10^{14} per cm^3), coagulation occurred simultaneously with nucleation and condensation growth of particles. Coagulation of very fine particles can involve the coalescence or sintering even at relatively low temperatures, and leads to the formation of particles of continuous structure rather than aggregates. With Δt intervals small enough, the changes caused in each of these intervals by any one of the three simultaneously occurring processes—nucleation, condensation growth, and coagulation—are relatively small, and these processes can be considered to run independently, preceding one another in each time interval.

Several procedures have been described in the literature[7-11] for calculating coagulation of polydisperse aerosols based on the solution of a set non-linear differential equations of the form (1.4) but without the terms taking account of particle evaporation.

By means of modern computers it is possible to solve a system of no more than a few hundreds of such equations. Therefore these methods are applicable only to processes in which the particles being formed contain no more than 100–150 primary particles. In the case of HDA formation, primary particles are single molecules, and the minimum value of i to commence calculation from is 1. For g this minimum value is g^*+1; at $g < g^*+1$, the concentrations are considered to be equal to equilibrium ones at each given moment.

Since particles of size 100 Å, i.e. consisting of 10^5 and more molecules, can be formed as the result of the process, the above-mentioned procedure is inapplicable for calculations in the case of formation of not very highly dispersed aerosols. Rosinski and Snow[12] proposed a simplified program based on the assumption that all particles consist of 2^i molecules, where i is an integer. A collision of two particles of the class 2^i leads to the formation of a particle of the class 2^{i+1}. Further, it is conditionally assumed that a collision of particles of the classes 2^i and 2^{i-1} also results in the formation of particles of the class 2^{i+1}, but in the quantity $(2^i+2^{i-1})/2^i = 0.75$ of the number of particles of the class 2^i participating in collisions. In the case of collisions of particles of the classes 2^i and 2^{i-j}, $j > 1$, the size of the former does not change, but their number increases $(2^i+2^{i-j})/2^i$ times. The first of these assumptions gives a too high value for the mass transfer rate along the particle size axis and the second a too low one, so that the errors due to these inaccuracies are at least partially compensated.

In reference 6 this program was combined with the scheme of calculating nucleation and condensation growth described above. Both calculated data and experimental results obtained by electronic microscopy of particles sampled at various stages of aerosol formation, in addition to the results of the determination of the sintering capacity of powders obtained by deposition of these aerosols, show that the process of silver HDA formation in this concentration range can be divided into six stages: (1) Nucleation and condensation growth. (2) Nucleation, condensation growth, and coagulation involving coalescence of particles. (3) Coagulation with coalescence and condensation growth. (4) Coagulation with coalescence. (5) Coagulation involving coalescence of fine particles and sintering of large ones. (6) Coagulation leading to formation of aggregates of particles held together only by cohesion forces. The particle size in aerosols obtained in this study, calculated from the specific area of deposits, was determined almost exclusively by coagulation involving coalescence of particles, and hence the mean particle size was proportional to the cubic root of vapor concentration.

Very high vapor supersaturations can lead to the establishment of a labile state, i.e. the thermodynamic barrier between the phases can disappear, and in that case the formation of clusters, even of two molecules, becomes an irreversible process. The assumption of the establishment of the labile state was arbitrarily made by Rosinski and Snow[12] in calculating the condensation of Fe_2O_3 vapors and by Stockham[13] for silver vapors. Calculating the condensation of silver vapors under the conditions similar to the Stockham experimental conditions, Sutugin and Fuchs[6] did not find a labile state to be established in that case. However, since in the final analysis the dispersion degree is determined by coagulation accompanied by particle coalescence and coagulation is of an asymptotic nature, the condensation behavior at early stages of aerosol formation has but little effect on the final result. For this reason, the Rosinski–Snow scheme is applicable also to the cases where no labile state is established, and accounts for the fact why Stockham obtained reasonable results.

At $g^* = 2$ the nucleation rate is determined by the ratio between the rates of trimer formation and dimer dissociation. The formation of the bound state in the case of a collision of two molecules was considered in reference 16. But for the life of a twin cluster formed depending on temperature, the value of the interaction potential and geometric collision parameters could be determined only by numerical methods. Therefore the results of reference 16 are as yet inapplicable for calculations of the aerosol formation process. It should be noted that the concentration of dimers in nitrogen calculated by this method differs somewhat from that calculated by Reed's method.[6]

In the cases where g is large enough and supersaturation varies but little during the relaxation time of the nucleation process, as given by the solution of a time-dependent diffusion equation (1.2), it is possible to make use of a conventional method of step-by-step calculation in which the nucleation rate at each step is determined by steady-state solutions of the classical theory equations. Naturally, this method is applicable neither in the presence of coagulation nor for description of the condensation of associated vapors. Using this method, Griffin and Sherman[15] showed that HDAs can be formed during condensation of copper vapors issuing from a Laval nozzle at large Mach numbers. A detailed description of step-by-step calculations of condensation in nozzles with the use of steady-state solutions of the classical theory equations is to be found in references 16 and 17.

Under the same condition of slowly changing supersaturation, the condensation process can be described by means of differential equations expressing the changes with time of temperature, vapor concentration, and nucleation rate, and—in the case of condensation in nozzles—also by equations of continuity and adiabatic expansion. Using a number of assumptions, the solution of a system of these equations in the simplest cases can be carried out analytically, i.e. a theoretical time dependence can be obtained for the number of particles formed and their mass. Such solutions were given by a number of authors.[18–20] An attempt to avoid simplifying assumptions was made in reference 21 yet necessitated the use of numerical methods of solution by means of computers.

Very often the formation of condensation HDAs involves chemical reactions in the gas phase, as, for example, in production of carbon black by pyrolysis of hydrocarbons and metals by decomposition of chlorides and carbonyls, etc. It is possible to distinguish three cases of formation of condensation aerosols due to chemical processes. In the first case the chemical reaction leading to the formation of condensing vapor occurs so fast that it is over before the nucleation starts. In this case the calculation of the condensation process is the same as that of "physical" condensation. In the second case the reaction runs simultaneously with condensation, but the aerosol particles formed have no effect on the reaction course. Here the chemical reaction can be taken into account by the step-by-step calculation of condensation. For this purpose it is necessary to take into consideration not only the vapor consumption in the growth and formation of particles, but also the formation of vapor as the result of chemical reaction as well as, if necessary, the thermal effect of the reaction. A program of such a calculation was described by Dunham,[22] who determined the dependence between the rate of H_2SO_4 vapor formation and the nucleation rate in this vapor. The third case, where the chemical reaction can be autocatalyzed at the surface of the particles formed, presents great difficulties, and no mathematical model of such process has been proposed so far. It should be mentioned that a method has been suggested by Tessner[23] for the calculation of the number of carbon black particles formed by pyrolysis of hydrocarbon vapors, as a function of time, if the particle size distribution in the final aerosol is known.

1.2. Generation of HDAs by physical condensation of vapor

The methods of generation of condensation HDAs can be classified according to the mode of producing vapor supersaturation. In physical condensation, supersaturation is produced by cooling vapor or vapor–gas mixture; in chemical condensation a substance is formed as the result of a gas reaction, with a very low vapor pressure at the reaction temperature. Vapor cooling can be effected by mixing it with cold gas, by heat exchange with a cold body (condenser walls, cooling agents, etc.), by adiabatic expansion, or by combination of these procedures.

The simplest, but poorly controllable method of HDA generation by mixing, consists in heating high-boiling solids or liquids in a turbulent stream of cold gas. In this case the mixing of the vapor of the substance being heated and gas, and the initial nucleation stage, occur near the surface of the evaporating body. In the absence of flow, or in the case of a laminar gas flow, mixing occurs very slowly and, as stated in section 1.1, only coarse aerosols are formed as a rule. Great difficulties involved in this method are due to the presence of more volatile impurities in the substance being heated, which also form HDAs upon heating. In most cases, however, these aerosols can be distinguished from the basic aerosol since their concentration decreases gradually as the impurity evaporates.[24] When the aerosol concentration remains constant over a long period of time, it can be considered to consist of the basic substance. This method is often used in a different form, viz. a thin layer of the evaporating substance is spread over a high-melting metal wire heated by electric current. Thus, for example, HDAs are obtained from various salts. A serious drawback of this technique is that the concentration of the aerosol being formed and the particle size decrease steadily with time. In the case of sodium chloride at 450°, the particle size decreases over the period of 5–7 hr.[25] This is accounted for by gradual recrystallization of the salt upon heating, leading to a decrease of its evaporation rate. In order to obtain constant sodium chloride aerosols, it is necessary to train the sodium chloride layer for several hours and to renew it after a day's operation: HDAs thus obtained are rather polydisperse. HDAs can also be generated by heating pure metal wires, mostly platinum, tungsten, and nickel, in air. The concentration of HDAs formed is rather constant in time and more or less reproducible and, on this account, they are often used as condensation nuclei in generation of monodisperse aerosols. The mechanism of formation of HDAs from the metals mentioned seems to consist in their surface oxidation and evaporation of oxides, which are much more volatile than the metals themselves. This follows from the fact that the formation of oxide HDAs starts at the temperatures at which the pressure of metal vapors is infinitesimally low: from platinum at 200°[26] or 255°C,[27] from tungsten at 300°C,[28] from nickel at 650° or 950°C.[24] Such large discrepancies in the values of the starting temperatures of HDA formation are due to the difficulty of detecting aerosol particles of almost atomic size and to the presence of hardly removable impurities in metals. These aerosols are also very polydisperse.[27, 29] If air was passed at the rate of

1 l/min around a platinum wire heated to 300°C, the aerosol concentration was 2×10^5 per cm^3, and the mean particle size $\approx$ 40 Å.[28] A nichrome wite heated to 1000°C produced a CrO_3 aerosol of particle size <15 Å.[29]

A similar method consists in sublimation of a substance in rarefied gas: the substance contained in a vessel with rarefied inert gas is heated either by a tungsten coil,[30–32] or by induction current.[33,34] This method was first proposed by Gen, Ziskin, and Petrov[30] for generation of aluminum aerosols, to be used later for a great variety of metals. The size of particles generated by this method decreases with decreasing metal evaporation rate and gas pressure, the minimum mean particle size obtained being approximately 200 Å. It is impossible to obtain more highly dispersed aerosols, since at low vapor concentrations the particle size is practically independent of gas pressure and vapor concentration. Turkevitch[34] studied in detail the mechanism of aerosol formation by this method. By locating sampling devices of aerosol particles at varying distance from the vapor source and using an electronic microscope, he succeeded in following the process of aerosol formation, or at least its last stages, and arrived at the conclusion that aerosol formation occurs in three steps, corresponding to steps 1, 3, and 5, described on page 8. Turkevitch attributed the particle-size decrease with lowering inert gas pressure to a greater rate of vapor diffusion and ensuing increase of its cooling rate.

In reference 13 a flow line version of this method was used: silver vapors from a crucible heated by induction were mixed with a rarefied argon stream flowing around the crucible. Thus aerosol vapors of mean particle size greater than or equal to 40 Å could be obtained. The particle size increased with vapor concentration, but was practically independent of argon pressure. In this case, vapor mixing with cold gas seems to be mainly caused by turbulence. Aerosol generation in rarefied vapors is of great use in studying aerosol properties at low pressure, since the problem of changing the pressure in aerosol without loss of particles, their coagulation, etc., is very complicated. However, for the same reason this method is inconvenient for aerosol investigations under normal conditions.

More refined are HDA generators in which a relatively large amount of substance is heated in a thermostated oven in a slow gas stream. Here gas is practically saturated with the substance vapor and then quickly mixed with cold gas. Under these conditions, vapor concentration remains practically constant with time, which favors constancy of the concentration and of dispersion degree of the aerosol formed. Two versions of these generators are in use. In one of them, mixing is effected in a free space e.g. in a hot, turbulent jet of vapor–gas mixture issuing into a large volume of cold gas. In this case, all vapor is condensed in the space. In the second version, mixing occurs in a T-piece, a nozzle with exit cone, etc., and a major portion of vapor is condensed on the mixer walls.

A generator of the former type was used in reference 6. The dependence of the dispersion degree of HDAs on the operating conditions of such a generator was studied for the case of an aerosol obtained by condensation of MoO_3 vapor contained in a hot-air jet issuing into cold air at a speed of 8–580 m/sec.[35] In the range of vapor concentrations in the jet of 0.1–100 g/m^3 it was possible to obtain aerosols of mean particle size 70–1000 Å. The particle size decreased down to a certain limit with decreasing vapor concentration and increasing flow velocity, and, if the latter was constant, with decreasing nozzle diameter. It is of interest to note that the particle size in aerosols formed in the case of condensation in a jet was smaller than that of aerosols obtained in a heat exchange generator (see below) at the same vapor concentration. The effect of the flow velocity and nozzle diameter was accounted for by means of an equation deduced in reference 6 for the rate of mixing in a free jet, which shows this rate to increase with the above-mentioned change of these parameters. In the concentration range of less than 1 g/m^3 the particle size increase with rising mass concentration was not very significant at first, becoming greater later, which can be due to a more decisive role of coagulation accompanied by sintering of particles.

The formation of HDAs in arc and plasma burners is a similar process. The substance is introduced either into the electrode material or as powder into the gas jet. The very hot vapor jet is cooled by mixing with cold gas or an atomized refrigerant. The vapor condensation upon cooling seems to occur on ions. According to the electronic microscopy data,[34, 36–37] the particle size of aerosols thus obtained is 25–1000 Å and the mass median size can be varied within 200–800 Å. By vaporizing metals in an inert gas or in a reducing one, it is possible to obtain metal aerosols, and, in an oxygen-containing atmosphere, oxide aerosols. It is of interest to note that the particles of aerosols generated by means of arc or plasma burners are usually amorphous. Holmgren, Gibson, and Sheer[37] showed that it is possible to increase the dispersion degree of aerosols by forced cooling of gases issuing from the arc. Selover[38] studied the formation of nickel aerosol upon cooling of argon plasma containing nickel vapors. Atomized nickel carbonyl was introduced into argon fed to the plasma burner. Since carbonyl decomposition occurred at a very high temperature, it could be considered to precede physical condensation. In the carbonyl concentration range 0.006–0.25 g/l, the particle size of the aerosol obtained varied within 15–150 Å. For rapid cooling the argon jet was sprayed with cooled isopropyl alcohol. With the mean particle diameter determined by the BET method equal to 200 Å, practically all particles were of the size within the 30–300 Å range.

Sofronov[39] obtained a highly dispersed carbon-black aerosol by condensing carbon vapors formed during decomposition of benzene vapors in plasma.

The plasma method makes it possible to obtain aerosols from substances with very high boiling points (with the exception of thermally unstable ones) and in large amounts. Unfortunately, the mechanism of the processes involved in condensation of cooled plasma jets has been insufficiently studied.

Now we shall pass to generators in which mixing of vapor and gas occurs in narrow mixers. In reference 40 a mixing generator is described for generation of HDAs from salts, such as sodium chloride, which consists of a quartz tube placed in an electric oven. One side of this tube ends in a T-piece of internal diameter 8 mm. Ceramic rings covered with the substance being evaporated are placed into this tube and air is passed through it at the rate of 0.1–0.5 l/min. In the T-piece the air containing salt vapor is mixed with a large volume of cold air (5–20 l/min). To prevent vapor condensation on the walls before mixing, the generator is provided with a superheater. In this generator it is possible to obtain monodisperse sodium chloride aerosols of the mean particle radius 14–100 Å. The maximum particle concentration that can be achieved increases with their size, and lies within 10^3–10^9 per cm^3. By varying four parameters—the velocities of the air flow through the furnace and through the T-piece as well as the current in the coils of the oven and of the superheater—it is possible to vary the mean size and concentration of aerosol particles in the range indicated. The generator is steady in operation and gives fairly reproducible results. A generator of this kind was successfully used in reference 41 for generation of radioactive HDAs.

A generation of reproducible monodisperse aerosols of dioctylsebacate of mean particle radius 30–300 Å is described in reference 42. These aerosols were obtained in the PSM described later in section 2.1, NaCl particles of mean size of 34 Å being used as nuclei.

The size of dioctylsebacate particles could be easily varied by changing its evaporation temperature, condensation nuclei concentration, and the ratio of flow velocities of the vapor–gas mixture and cold, diluting air.

The vapor–gas mixture can also be cooled by heat exchange with the generator walls. This method was made use of in the well-known Sinclair–La Mer generator of monodisperse aerosols. La Mer[43] pointed to the possibility of using this generator to obtain HDAs of sulfuric acid. According to La Mer, it is possible to generate by this method HDAs of particle diameter 20 Å.

Matijević and Kerker with collaborators[44–46] modified La Mer's generator and used it for generation of aerosols from various mineral salts—AgCl, NaF, and NaCl. They studied in detail the effect of the gas flow velocity in the generator and of the evaporator temperature on the dispersion degree of the aerosol formed under the conditions of spontaneous condensation and found the mean particle size to rise with increase of these parameters. The reason of the particle size increase with rising flow velocity is not quite clear. Unfortunately, no detailed scheme of the modified generator has been published. Particles of the size from 10 to 1000 Å could be obtained, the degree of polydispersity of aerosols depending little on the mean particle size and the standard geometric deviation being 1.15–1.2 Å.

In a number of studies, generators of the heat-exchange type were used for generation of sodium chloride of HDAs with high weight concentrations (up to 1 g/m^3) in order to obtain high-dispersed powders.[47–49] The authors of these studies also observed particle size increase with higher vapor concentrations. Heat exchange does not allow very high cooling rates, and, hence, high vapor supersaturations. Therefore in heat-exchange generators it is difficult to obtain aerosols of particle size less than 100 Å.

In reference 35 MoO_3 aerosols were generated by passing a hot air stream containing MoO_3 vapor at the ratio of 1–28 l/min through a water-cooled quartz tube 20 mm in diameter. At vapor concentrations 0.002–8 g/m^3, the mean particle size, determined from the surface area measurements (by the BET method) of the powder obtained by precipitation of these aerosols, was 130–1000 Å. The flow velocity was not found to have a marked effect on the dispersion degree of aerosols.

Besides heat exchange and mixing with cold gas, cooling can be effected by adiabatic expansion. By this process vapor condensation occurs in the nozzles of supersonic wind tunnels where it is an objectionable phenomenon to be eliminated by special measures (air drying, heating of nozzle walls, etc.). On the other hand, formation of HDAs—especially strongly charged ones—in the nozzles is of great interest for the electric propulsion development.

For all substances except helium the decrease of the equilibrium pressure of vapor due to the temperature drop in adiabatic expansion occurs faster than the static pressure decrease. Therefore a large enough isentropic expansion of the vapor–gas mixture, or even of pure gas, can lead to high supersaturations. The rate of the supersaturation increase can be very great, and, hence, in wind tunnels HDAs can be formed even of such highly volatile substances as water. The lifetime of these aerosols, however, is very short, the particles evaporating with increasing temperature in a compression shock. In references 50 and 51 the size of small ice crystals formed in a Laval nozzle at Mach numbers $M = 5$ was estimated from their inertial properties and evaporation rate in the compression shock. The particle size was approximately 14 Å, being about the same for all particles and independent of water vapor concentration.

In references 52 and 53 the condensation in Laval nozzles was studied in connection with the possibility of obtaining a beam of charged particles to be used in an electric thruster. Hg_2Cl_2, $HgCl_2$, and $AlCl_3$ vapors were passed through a heated nozzle into a vacuum.(52) For examination of particles they were deposited on a net placed in the nozzle. This method gave particles of mean size 100–1000 Å. In reference 53 cadmium vapors issuing from a nozzle were condensed in a positive corona discharge field, the particles obtained being of the size, approximately 100 Å.

The cooling of the vapor–gas mixture both by adiabatic expansion by mixing with cold gas and by radiation seems also to take place in aerosol generation by the exploding-wires method.(54–56) In this method a high voltage discharge of considerable strength (about 1 kJ per 1 mg of wire substance) is applied to a thin wire held between massive electrodes. The wire instantaneously changes to a plasma pinch, the cooling of which causes the formation of an aerosol. Metal aerosols are formed in an inert gas from all metals and in air only from noble metals. In the presence of nitrogen and oxygen, nitrides and oxides are formed of particle size within 50–1000 Å. The mean particle size depends on the wire material and decreases with increasing discharge energy. The fact that instantaneous dilution of the aerosol cloud gives aerosol consisting of individual unaggregated spheres(34) shows that in aerosol formation of prime importance is the condensation growth rather than coagulation. The particle concentration in the cloud formed during explosion being rather large (approx. 10^{11} per cm^3), some coagulation does occur, but in an already cooled aerosol, so that it does not involve coalescence of particles.

In reference 51 an aerosol generator is described in which the cooling of the vapor–gas stream is effected by means of two processes—adiabatic expansion and mixing with cold gas. In this generator a hot vapor–gas jet flows with a great velocity through the nozzle into cold gas. This method gave, without use of nuclei, rather monodisperse selenium aerosols of particle size 600–2600 Å.

1.3. Generation of HDAs by chemical reactions in the gas phase

Such processes as thermal and photochemical decomposition of carbonyl vapors, thermal hydrolysis of metal chlorides and silicon tetrachloride vapors, and thermal destruction of hydrocarbons have been used for a long time on an industrial scale for production, through an aerosol stage, of highly dispersed powders of metals and their oxides, carbon black, and aerosol. There is a voluminous patent literature on all these processes, the consideration of which is beyond the scope of this article, so only a few studies, where the attention was centered on the mechanism of aerosol formation, will be discussed. The particle size in aerosols from metals and their oxides obtained in the studies discussed below was determined from the surface area of deposits measured by the BET method.

The photolysis of iron pentacarbonyl vapors in the presence of oxygen was repeatedly used for aerosol generation.(57–61) It was found in reference 62 that one light quantum causes one carbon monoxide molecule to split from the carbonyl complex, followed by the formation of a $Fe_2(CO)_9$ complex, which is oxidized to Fe_2O_3 in the presence of oxygen. In reference 63 this reaction was used to obtain the molecular condensation nuclei. The process of Fe_2O_3 aerosol formation was studied in detail in reference 64. It was found that with increasing illumination time (from 0.1 to 6 sec) a mercury lamp of a mixture of nitrogen with $Fe(CO)_5$ vapors, the decomposition extent of the latter increased from 10 to 100%. After illumination the unreacted carbonyl was adsorbed by passing the aerosol through an activated carbon layer which diminished the effect of autocatalytic decomposition of carbonyl on the surface of the particle. Using a carbonyl concentration of 36.2 g/m^3 with decomposition extent 10–100%, aerosols of mean particle size 22–82 Å could be obtained.

The generation of aerosols of metal oxides by thermal hydrolysis of the vapors of their chlorides was studied by a group of French authors [65, 66] who fed a hydrogen–oxygen burner with these vapors. The chloride concentration was 3–12 mg/l and the oxygen–hydrogen flame temperature 1600–3000°K. The mean particle size of HDA deposits was 100–1200 Å, and decreased with decreasing vapor concentration. The flame with temperature of 1900°K gave more highly dispersed aerosols than at 1500°K or at 3000°K. At high temperatures the particle size is increased, owing, probably, to sintering of particles upon coagulation, whereas at too low temperatures slow hydrolysis, and hence low nucleation rate, results. In the flame of a larger burner more coarse particles were formed which was accounted for by the cooling rate of the free jet decreasing with increase of its diameter.

In reference 67 thermal hydrolysis of $AlCl_3$ vapors was studied by mixing the nitrogen stream containing these vapors with air saturated with water vapor. Hydrolysis occurred at 400–450°C. More highly dispersed aerosols were formed in this case than in the flame. At chloride concentrations of 5–30 mg/l, the mean particle size was 50–400 Å. Evidently low temperatures used in reference 67 eliminated sintering of particles and were conducive to a better dispersion degree of the deposit than in references 65 and 66.

A review of the papers on the production of carbon black and Aerosil can be found in references 68 and 69 and the formation of HDAs during radiolysis of SO_2 is treated in reference 70.

In reference 71 HDAs of sulfuric acid were generated by mixing air streams containing SO_3 and H_2O in a capillary T-piece. The investigation of the particle size of these aerosols is described in section 2.2.

1.4. Generation of HDAs by atomization

Atomization of liquids involves the formation of particles of a great variety of sizes, including very fine ones. However, true HDAs containing no coarse particles (or only a few) cannot be obtained by atomization, nor is it possible to obtain HDAs by atomization of highly dispersed powders since, owing to strong adhesion between very fine particles, complete disaggregation of these powders cannot be achieved. However, by atomizing very dilute solutions and drying droplets, one can obtain HDAs of the solutes. In reference 72 this method was used for generation of aerosols from various salts of mean particle size 100–1000 Å. Aerosols of mean particle size 600 Å were generated by atomizing salt solutions having concentrations equivalent to 0.1% and separating large drops by bubbling followed by drying of droplets.[73] A modified La Mer generator, known as the Rapaport–Weinstock generator, was used to generate monodisperse dioctylphthalate particles of 400 Å size. The mist obtained by mechanical atomization of a dioctylphthalate solution in methanol was passed through a heater, with both components evaporating.[74] Then the vapor–gas mixture was passed through an air cooler, the methanol vapors either condensing on the walls or remaining in the vapor phase, whereas dioctylphthalate vapors condensed both on the walls and on the non-voltaile residues of evaporated particles acting as nuclei. The size of the particles thus obtained can be varied by changing dioctylphthalate concentration in the solution.

The generation of a monodisperse aerosol of particle diameter 520 Å by atomization of a virus culture is described in reference 75.

In connection with the progress made in the development of colloid thrusters and transformers of energy, experiments have been reported on atomizing liquid cesium by a strong electric field in a vacuum which gave particles of the size greater than or equal to 40 Å.[76] Although these particles are suspended in rarefied cesium vapors rather than in a gaseous medium, it is possible to consider this system as aerosols. It is interesting to note that electrostatic atomization of liquids at normal pressure does not give particles less than 2000 Å.[77]

CHAPTER 2

METHODS OF INVESTIGATION OF HDAs

2.1. Investigation of HDAs in suspended state

Inasmuch as the properties of aerosols of particle size less than 1000 Å differ from those with larger particles, methods of investigation are, of necessity, different also. There are two main groups of these methods—those using aerosols in suspended state and those based on the measurement of particle deposition rate or on the examination of deposits. The minimum size of particles that are visible in the ultramicroscope under normal illumination conditions are 1000 Å, and the light-scattering intensity in HDAs decreases very rapidly with particle size according to Rayleigh's law. HDAs possess a very low optical density unless their concentration is very great. In the latter case, however, they coagulate very rapidly. Therefore, to be able to use optical methods for the study of HDAs, the particles should be enlarged by condensing on them vapors of some substances. For instance, in the adiabatic particle counters of Aitken and Scholtz,[78] which have long been used by geophysicists for counting atmospheric condensation nuclei, the particle size is increased by condensing water vapor on them. The supersaturation produced in such counters by adiabatic expansion is usually sufficient for non-hygroscopic particles larger than 100 Å to become visible.

It is very difficult to determine accurately the concentration of particles of a still smaller size, even using increased supersaturation, owing to diffusional deposition of particles on the walls of the counter chamber and of the entrance channel during sampling and in the time interval before expansion. Nolan and Pollack[79–81] have devised a photoelectric adiabatic counter in which the mist formed does not settle on the bottom as in Aitken's counter, and the particle concentration is determined from light scattering or extinction in the mist, measured by means of a photoelectric cell. Initially, the counters of this type were calibrated by means of an Aitken's counter, and thus all the errors due to it were also implemented in the automatic counter.[81,82] A counter used for calibration is described in which the concentration of enlarged particles is determined by stereomicrophotography in suspended state. In this counter some of the drawbacks of Aitken's counter were eliminated, although the chief source of errors—the diffusion losses of very fine particles—remained. The early versions of photoelectric counters had poor reproducibility, and readings showed a tendency to drift with time.[83] Later, these defects were eliminated.[84] It should be noted that the calibration curve of a counter depends on temperature and pressure.[85] The optical density of the mist[85,88] is related to its particle concentration n by an empirical formula $E \sim n^{0.46} M^{0.54}$, where M is the mass of vapor condensed in unit volume. Some other instruments based on this principle were described by Verszar[86] and Rich.[87]

Automatic adiabatic counters have been described in which consecutive expansions are effected at a considerable rate, as many as five times a second.[89,90] In this case, the residence time of aerosol in the chamber and entrance channel is reduced, which should lead to decreased diffusion losses. However, such high rate of expansion necessarily leads to turbulization of the aerosol stream on its way to the chamber, which can increase the losses. In conclusion, it should be pointed out that on account of diffusion losses, adiabatic counters can hardly be used for investigation of particles of less than 100 Å.

Flow counters have been described in which diffusion chambers are used to produce supersaturation.[91–93] However, the supersaturation achieved in these counters is very small, sufficient only for revealing rather large particles, of $d > 400$ Å; the aerosol flow velocities in these chambers are very low.

An ingenious flow method for enlargement of liquid particles was proposed by La Mer.[43,94–95] The mists of dibutylphthalate and sulfuric acid were passed over toluene and water respectively. In a very short time the composition of the droplets comes into equilibrium with the vapor phase, and the enlarged droplets are studied by optical methods. By means of simple calculations it is possible to find, in addition to the particle concentration, also the size of original droplets.

In reference 96 the condensation of water vapor on the particles was achieved by mixing HDAs with a hot vapor–gas mixture.

A flow method for enlargement of HDA particles suggested by Kogan and Burnasheva[97] proved to be very successful. In their particle size magnifier (PSM) (Fig. 1) the stream of cold HDAs is mixed in a diffusing cone with a hot gas stream containing vapors of a high-boiling liquid, such as dibutylphthalate. The streams enter the cone coaxially—the hot vapor–gas mixture in the center and HDAs along the periphery. Turbulent mixing of the streams produces supersaturation whose degree can be controlled by the streams ratio, flow velocities, temperature in the vaporizer, width of the annular gap through which HDAs enter into the diffusing cone, as well as by the choice of the working fluid. Thus a stable monodisperse aerosol of mean particle size some tenths of a micron is generated which can be examined by conventional optical methods. A remarkable feature of the PSM is the independence of the enlarged particle size of their concentration. This is accounted for by most of the vapor condensing on the diffusing cone walls rather than in the volume, this surface condensation determining mainly the actual vapor supersaturation in PSM and, hence, the size of the particles formed. In the improved model of the PSM[63] the upper limit of particle concentration at which it does not affect the size of enlarged particles is equivalent to 10^5 per cm^3. Due to this feature, it is possible to determine relative HDA concentrations nephelometrically without preliminary calibration. Using the PSM with dioctylsebacate vapor, Kogan succeeded in developing even single molecules of some substances.[63]

The measurements of light scattering by HDAs entering the PSM and issuing from it enables determination of the mean size of particles greater than 400 Å if the particle concentration of HDAs is large enough to give appreciable light scattering.[97]

In reference 98 the losses of particles in the PSM were studied. For this purpose, a version of the PSM was designed in which HDAs were introduced into the diffusing cone along the axis and not along the periphery, as in the original PSM, whereas a hot vapor–gas mixture moved along the walls. This eliminated the diffusion losses. By comparing the concentrations of aerosols obtained in both versions of the PSM, it was shown that the losses of sodium chloride particles of $d = 50$, 90, and 200 Å in the original PSM were

33, 17, and 10% respectively. On the other hand, the modified PSM gave less monodisperse aerosols, and the particle size was somewhat dependent on their concentration.

Aitken[99] was the first to measure the activity spectrum of the condensation nuclei, and Nolan[100] pointed out that the supersaturation necessary for the development of particles in his counter can be also used for their size determination. Later, in reference 101, this method was used to estimate the size of the sodium chloride particle. To produce low vapor

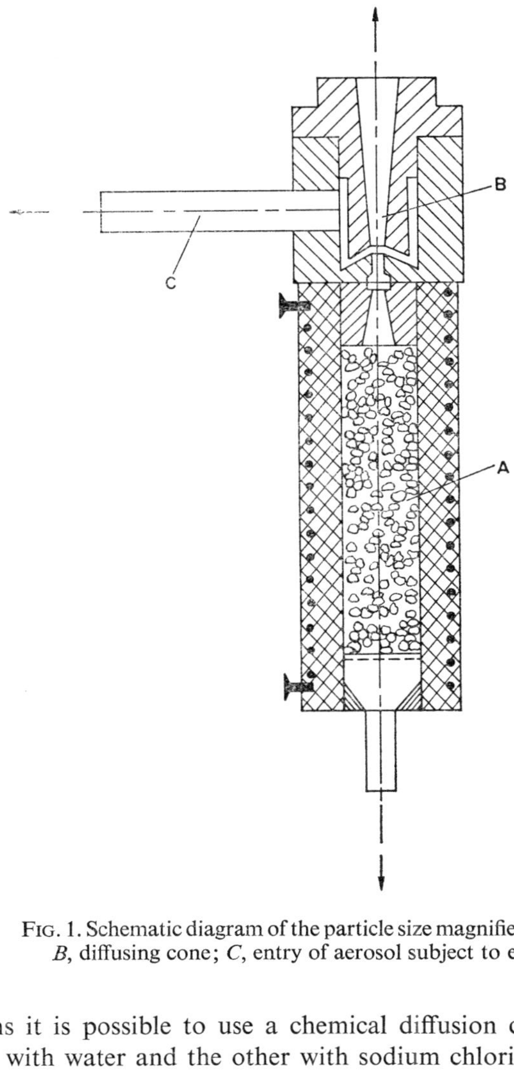

FIG. 1. Schematic diagram of the particle size magnifier. *A*, vaporizer; *B*, diffusing cone; *C*, entry of aerosol subject to enlargement.

supersaturations it is possible to use a chemical diffusion chamber in which one of the walls is wetted with water and the other with sodium chloride solution.[102] By means of this chamber it is possible to produce supersaturations in the range of 0.01–1% within 0.01%. It was shown in reference 103 that the relative humidity at which the enlargement of various mineral salt particles of $d = 200$ Å starts, is close to that calculated by means of well-known thermodynamic equations. Evidently, this method can be used to determine the mean size, or even the particle size distribution, in aerosols consisting of one definite substance with known wettability and hydroscopic properties. However, this is impossible for aerosols with particles of an unknown composition.[104] The theory of development of aerosol particles in a diffusion chamber was published by Buykov.[105–106]

In reference 107 a method was suggested for calculating the size of uncharged particles from the coefficient of capture of gas ions by them. This coefficient can be found by measuring the decrease of ion concentration upon introduction of the aerosol being examined into a bipolar ionic atmosphere. This requires, however, an independent determination of the aerosol particle concentration. If we take into consideration also the fact that the dependence of the capture coefficient on the particle size has not been accurately determined as yet, the possibilities of this method seem to be rather doubtful.

For determination of particle concentration and size of substances, whose vapors are intensively luminescent at high temperatures, such as sodium or lithium salts, a flame counter was proposed[108–109] in which aerosol is introduced into the colorless flame of a hydrogen burner. The passage of aerosol particles is attended by flashes which are recorded by a photomultiplier with a pulse discriminator. A version of this counter described in reference 110 permits measuring the number and size of sodium chloride particles of $d > 400$ Å. The upper limit of the particle concentration determinable by the counter is rather low on account of the error due to coincidences.

Recently the small-angle scattering of X-rays has been used for determination of the particle size distribution in high-dispersed suspensions. It can be expected that in the future this method will be used for concentrated HDAs as well, especially for investigating the processes of their formation under steady-state conditions.

2.2. Methods of investigation of HDAs based on particle precipitation

This group includes a large variety of methods. Let us consider first the methods of determination of the electric mobility of particles, i.e. the velocity of their motion in an electric field of unit strength. A classical method for determination of electric mobility (and concentration) of gas ions by passing ionized gas through a condenser and measuring the current flowing through it as the function of applied voltage, was first used by Langevin[111] for investigation of heavy ions, i.e. charged particles of atmospheric aerosols. Since HDAs of $d < 600$ Å contain practically no particles with charges greater than one elementary charge (see p. 46), it is possible to calculate from the electric mobilities spectrum the size distribution of charged particles. However, as the proportion of charged particles increases with their size, it is impossible by this method to find the size distribution of all particles of HDAs.

Thus, strictly speaking, this method is applicable only to monodisperse aerosols, but, on account of its simplicity, it is widely used for investigation of atmospheric aerosols. In reference 112 it was used to study HDAs of sulfuric acid. A considerable contribution to the development of the method in question was made by Israel.[113]

Similar to the above is a method in which HDA samples are introduced into the chamber of the Nolan–Pollak counter containing a flat electrode to which a potential difference is applied.[114] Then the concentration of particles remaining in the condenser after different time intervals (i.e. kinetics of the aerosol concentration decrease) is measured and, hence, the electric mobility distribution of particles, as well as the percentage of uncharged particles, is calculated. In this case the diffusional deposition of uncharged particles on the condenser plates has to be neglected, which, in the case of very fine particles, can lead to considerable errors.

By determining the percentage of uncharged particles in a more or less monodisperse HDAs charged in a bipolar ionic atmosphere, it is possible to calculate the particle size. As the first approximation, the Boltzmann equation can be used for this purpose. Then, for the uncharged particles' fraction, we obtain the expression

$$\frac{n_0}{n_0+n_1+n_2+\dots} = \frac{1}{1+2\exp[e^2/2akT]+2\exp[(2\epsilon)^2/2akT]}, \tag{2.1}$$

where n_0 is the concentration of uncharged fraction and n_1 that of particles with ± 1 elementary charge, etc. According to the most recent data,[115] an exponential equation, similar to the Boltzmann formula, describes correctly the charge distribution in HDAs of particle radius greater than 200 Å. For more highly dispersed aerosols the percentage of charged particles under steady-state conditions can be calculated theoretically by means of equations given in section 3.7. This method, proposed by Wright[116] and improved by Rich,[117] is applicable to particles of any size, but inapplicable to polydisperse aerosols. In the latter case, using eqn. (2.1), it is possible to calculate a certain equivalent radius a_e. As shown in reference 118, the ratio of a_e to the arithmetic mean radius is the greater the higher the aerosol polydispersity, and can serve as its measure.

It should be borne in mind that in using the methods based on the measurements of the steady-state distribution of electric charges that a certain period of time is required to achieve such a state; in particular, in the case of natural ionization it may take many hours.[119]

The diffusion method of determination of HDA particle size, first applied by Townsend to the measurement of the gas ions mobility, proved to be very fruitful. For aerosols it was first used by Nolan[120] and Radushkevich.[121] The method is based on measurement of the diffusional deposition of particles on the walls in the case of a laminar flow of aerosol through a channel of circular or square cross-section. The aerosol penetration, i.e. the ratio n/n_0 of concentrations at exit and entrance to the channel, can be accurately calculated theoretically, and from the penetration measurements it is possible to determine the particle diffusion coefficient D, related to their mobility B (velocity of motion under the action of constant force of 1 dyne) by the formula $D = kTB$ (for the dependence of B on the particle size, see section 3.2). The equations for the diffusional deposition (and for a mathematically equivalent problem—heat transfer in the channel) were obtained by a number of authors in the form of series.

For a cylindrical channel in the initial stage of diffusion ($\mu < 0.005$), when the particle concentration near the axis is still equal to the initial concentration, the penetration is expressed by the formula[122]

$$n/n_0 = 1-2.56\mu^{2/3}+1.2\mu+0.177\mu^{4/3}, \tag{2.2}$$

where $\mu = Dx/R^2\bar{v}$, x is the channel length, R is its radius, and $\bar{v}$ is the mean linear flow velocity. For a later stage ($\mu > 0.005$)[123]

$$n/n_0 = 0.819\exp(-3.657\,\mu)+0.097\exp(-22.3\,\mu)+0.032\exp(-5.7\,\mu). \tag{2.3}$$

In a plane-parallel channel with width $2h$, the penetration for both stages is accurately enough expressed by De Marcus' equation,[124]

$$n/n_0 = 0.9149\exp(-1.885\,\mu)+0.0592\exp(-22.3\,\mu)+0.0258\exp(-151.8\,\mu), \tag{2.4}$$

in which $\mu = Dx/h^2\bar{v}$.

A numerical integration of the diffusion equation for a plane-parallel channel, performed in reference 125 using an electronic computer, gave a result in fair agreement with the analytical solution obtained by De Marcus.

In practice, aerosol is often caused to flow through a system of channels—a device which Rodebush[126] proposed to call a "diffusion battery". In this case, the expression for μ should contain in the numerator a factor equal to the number of channels. The sedimentation rate of HDA particles is so low that there is no need for a vertical arrangement of batteries.

The equations given above are valid for deposition of a monodisperse aerosol, whereas practically all aerosols are polydisperse, and when flowing through a channel it is the finest most mobile particles that are deposited first. Therefore, the apparent diffusion coefficient of a polydisperse aerosol, which can be calculated by means of the equations given (see above), increases with increase of parameter $y = \mu/D$. Hence the diffusion measurements are of any value only if they give the distribution of diffusion coefficients, or at least the value of the coefficient corresponding to a definite mean particle size. Four methods have been proposed for the determination of the parameters of the particle size distribution from experimental diffusion data. Pollak and Metnieks[127] have suggested the so-called exhaustion method. If aerosol is considered to be a mixture of several monodisperse fractions, characterized by increasing values of the diffusion coefficients $D_1, D_2, \ldots, D_i$ and the particle concentration fractions $p_1, p_2, \ldots, p_i$, it can be shown that

$$\exp(-KD'/\bar{v}) = \sum_i p_i \exp(-KD_i/\bar{v}), \tag{2.5}$$

where D' is the empirical value of the diffusion coefficient determined at the flow velocity $\bar{v}$, and K is the coefficient depending on experimental conditions according to eqn. (2.3) or (2.4). At $\bar{v} \to \infty$, expanding into a series the exponentials in eqn.(2.5), we have $D' = \sum_i p_i D_i$, i.e. the weighted mean value of D' of the mixture of fractions. At $\bar{v} \to 0$ it is admissible to retain in eqn. (2.5) only the term corresponding to the fraction with the least D_i. Therefore, by extrapolating the experimental function $n/n_0 = f(y)$ to $\bar{v} \to 0$, we can determine D_1 and p_1. Subtracting p_1 from the total number of particles, we can calculate the form of the function $f(y)$ for an aerosol without the first fraction. Then we extrapolate again to $\bar{v} = 0$, find D_2 and P_2, etc.

This method was used by Irish scientists.[128,129] In eqn. (2.5) only the first term on the right-hand side of eqn. (2.4) was allowed for, as the second term leads to very cumbersome expressions. Nolan and Scott[130] studied the effect of the number of fractions on the accuracy of the method and found that a correct picture of the actual particle size distribution can be obtained by dividing aerosol into no less than nine fractions. However, in spite of cumbersome calculations, the accuracy of this method is not very good owing to the necessity of performing repeated extrapolations.

Fuchs, Stechkina, and Starosselski[131] calculated by means of an electronic computer the values of integral diffusion penetration versus $n/n_0 = f(y)$ for aerosols with log–normal particle size distribution $g(a)$ from the formula [see eq,. (2.4)].

$$n/n_0 = 0.9149 \int_0^\infty g(a) \exp[-1.885 D(a) y]\, da + 0.059 \int_0^\infty g(a) \exp[-22.33 D'(a) y]\, da + \ldots \tag{2.6}$$

Sets of $f(y)$ curves were obtained corresponding to a definite mean geometrical particle radius a_g in the range from 10 to 1000 Å and to the values of the logarithm of standard geometrical deviation β_g from 0 to 0.4. For the particle mobility, the Millikan formula was made use of (see section 3.2). It should be noted that the log–normal distribution was found to be valid for a great variety of aerosols. Figure 2 shows the $n/n_0 = f(y)$ curves taken from reference 131. In using this "method of best fit", in practice, the experimental data plotted in the coordinates $(n/n_0, \log y)$ are superimposed by the calculated curves plotted in the same scale on tracing paper, and the latter are shifted along the abscissa axis until the experimental points coincide most closely with one of the calculated curves. Thus β_g is determined. D is found from the shift along the log y-axis.

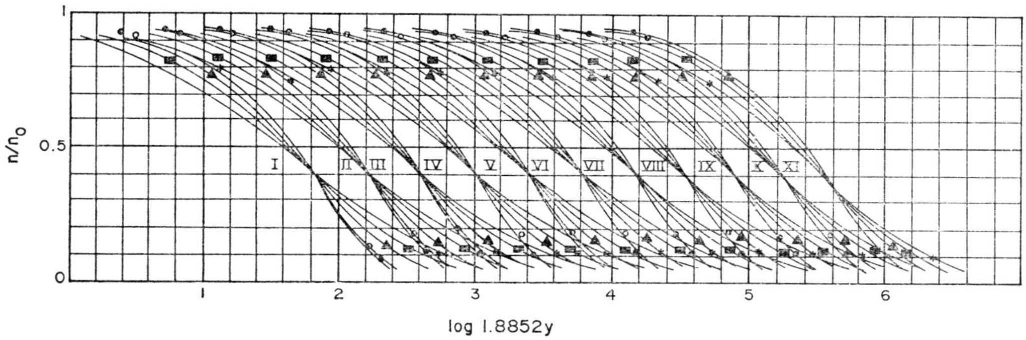

FIG. 2. Diffusional deposition of polydispersed aerosols in a plane-parallel channel. a_g(Å): I 10, II 16, III 25, IV 40, V 63, VI 100, VII 160, VIII 250, IX 400, X 630, XI 1000. Log β_g: ●—0, ○—0.1, ▲—0.2, ■—0.3, ×—0.4.

A most logical approach, which, however, proved to be unsuccessful in practice, was used by Twomey.[132] By transforming eqn. (2.3) or (2.4) for the case of polydisperse aerosols and obtaining an expression similar to eqn. (2.7), he showed that the experimental dependence $n/n_0 = f(y)$ can be presented as a converging series of functions of the form $g(a)A_i \exp(-\alpha_i y)$, which is a Laplace transform of the size distribution function $g(a)$ sought for. Therefore, this function can be found by means of the inverse Laplace transform. Theoretically, this method is applicable to aerosols with any size distribution. However, as was shown by calculations performed by Twomey himself by means of a computer, if we calculate $f(y)$ from $g(a)$ and then, by means of an inverse Laplace transform, change back from $f(y)$ to $g(a)$, the latter will be significantly different from the original $g(a)$. The error due to this circumstance increases greatly if, on account of imperfect measurements, the function $f(y)$ has been determined inaccurately. Only if penetration is determined over the whole range of measurements with the error $\leqslant 0.5\%$ is it possible to obtain reasonable results by this method. Such accuracy being usually unattainable, Twomey's method at the present seems to be applicable only for rough estimation of the particle size distribution in very polydisperse systems, e.g. in atmospheric aerosols.

Nolan and Scott[130] deduced an approximate formula for the parameters of the diffusion coefficients distribution

$$D' = \bar{D} - \sigma^2/2K\bar{v}. \tag{2.7}$$

Here $\bar{D}$ is the mean value, D' is the value found experimentally at the rate $\bar{v}$, and σ is the standard deviation of D. By determining D' for two values of $\bar{v}$, it is possible to find $\bar{D}$ and σ, and from them to pass to the mean particle size and standard size deviation. Recently, Metnieks[133] compared the three methods mentioned above (the method of inverse Laplace transform Metnieks did not consider to be practicable). The values of n-$n_0 = f(y)$

obtained by calculations from different particle size distributions were treated by the three methods. The calculations were performed for the distributions in which the number of particles is proportional to a^{-1}, a^{-3}, and e^{-a}, as well as for normal, log–normal, and a certain bimodal distribution. The curves calculated by the method of best fit[131] proved to approximate rather well the $f(y)$ dependences for all these distributions. The method of iteration also gave good results, but the Nolan–Scott method proved to be applicable only to normal and log–normal distributions, and even then its accuracy was less than that of the other two methods. Nolan and Scott[130] pointed out the advantages of the best-fit method over the method of iteration.

Recently, a fifth method of determination of the parameters of particle size distribution has been proposed, based on an asymptotic expansion of integrals in eqn. (2.6) with respect to the small parameter β_g.[163] By means of the analytical expressions obtained it is possible to find $\bar{D}$ and β_g from the measurements of D at two values of $\bar{v}$.

Equations (2.2)–(2.4) were obtained assuming the parabolic flow profile to be established from the very beginning of the channel. However, this profile is established, e.g. in a circular tube, only at the distance $0.05RRe$ from the entrance, where Re is the Reynolds number. A detailed consideration shows that the diffusional deposition in the entrance section is much larger than for a parabolic flow profile. The value of the diffusion parameter corresponding to the end of the entrance section being $\mu_e \approx 0.1D/\nu$, where ν is the kinematic viscosity of gas, the decrease of penetration through the battery due to this fact becomes larger with increasing diffusion coefficient. Practically, this effect is of importance only for very highly dispersed aerosols of particle radius $\leqslant 10$ Å. Thomas[134] attempted to eliminate the error due to the entrance effect by measuring the ratio of penetrations $(n_1/n_0)/(n_2/n_0)$ through two batteries, 47 and 5 cm long, and assuming this ratio to be equal to the corrected penetration through the 42 cm battery. However, this approach can lead to an appreciable error at small μ values due to the fact that penetration is not exactly expressed by an exponential function of the channel length. If the relative value of the second term in eqns. (2.3) and (2.4) is not very small, the product of the penetrations through two channels of the lengths l_1 and l_2 is not equal to the penetration through a channel of the length $(l_1 + l_2)$. The existence of this error was acknowledged by Thomas himself.[135]

Nevertheless, Twomey,[136] in order to obtain points corresponding to different y, proposed to pass aerosol through the same battery a varying number of times. This method leads to the same error, since the penetration through the battery of length L differs from the squared penetration through the battery of length $L/2$.

In reference 40 an empirical correction for the entrance effect has been proposed which can be used for investigation of aerosols with a small size distribution width. In the case of a monodisperse aerosol, the values of D calculated from experiment at various y should coincide. However, actually D depends on y even for a monodisperse aerosol, and this dependence is due to more intensive particle deposition in the entrance section compared to that in parabolic flow. If y is changed by using batteries of different length, rather than by changing the flow velocity through the battery, the relative influence of the entrance effect will diminish with increasing battery length, and its absolute value will be the same for measurements at all y. Therefore, the entrance effect can be estimated by comparing penetrations through batteries of different length. As seen from Fig. 2, the penetration curves for $\beta_g = 1.0$ and 1.26 at values of $n/n_0 \geqslant 0.45$ coincide. Thus it is possible to estimate the entrance effect also for aerosols of a certain polydispersity, but in this case penetrations over 0.45 should be compared. Recently, particle diffusion to the walls of a circular channel in the process of establishment of a parabolic profile was calculated by means of a computer.[(164)] Additional deposition of particles in the entrance section is 2% at the Schmidt number $Sc = \nu/D = 10$ and 5% at $Sc = 0.9$. This is in good agreement with the value of the empirical correction introduced in reference 40.

For the experimental determination of the function $f(y)$ the concentration of HDA before and after passing through the battery must be determined. For this purpose, Nolan and Pollak and collaborators used their adiabatic counter, the authors of this article,[40,42,98] —the PSM (see p. 18) in conjunction with a photoelectric nephelometer. Sometimes, when working with radioactive aerosols, for determination of n/n_0, radiography of filters is used through which the aerosols that have and have not passed through the battery, are drawn.[(138,139)] In reference 139 the ratio of activities of these filters was compared with the ratio of particle concentrations measured by means of an adiabatic counter before and after the battery. It was shown that the radiographic method gave 2–3 times lower values of D, which is understandable since the radioactivity of the aerosol particles activated by emanation decay products increases with their size. To determine the diffusion coefficient of the decay products of thoron, the activity distribution in the deposit on the walls of the diffusion channel was measured.[(140)] From the results of these measurements the values of the diffusion coefficient were calculated by means of the equation describing the deposition on the walls of the products formed in the diffusion channel itself.[(141)]

In designing the diffusion batteries it should be borne in mind that it is impossible to have a "universal" battery, applicable over the whole range of particle sizes of HDAs, since their diffusion coefficient varies from approximately 10^{-1} to 10^{-5} cm^2/sec. To obtain the required values of μ over the whole size range, it would be necessary to vary by four orders of magnitude the flow velocity through the battery, which is impossible since high velocities involve turbulent flow and low velocities—an error due to convection currents. Calculations and personal experience of the authors of this review show that for particles greater than 200 Å it is expedient to use plane-parallel batteries, and for smaller particles batteries of cylindrical channels. The authors used in their study of HDAs in the particle size range 14–60 Å a set of batteries of different length, consisting of eight 6 mm tubes, and for particles of 60–200 Å a set of sixty-channel batteries composed of 1.4 mm tubes. In general, it is advisable to use for batteries a large number of narrow tubes rather than several wide tubes, since in this case it is easier to widen the range of y-values of the battery by changing the flow rate. The use of a set of batteries of different length for determination of $f(y)$ eliminates errors in the flow rate measurement, since all measurements are performed at the same rate, and facilitates the introduction of a correction for the entrance effect.

As regards the sources of errors which may arise in using the diffusion method, in addition to the entrance effect already discussed, the acceleration of deposition of charged particles on the channel walls under the action of mirror forces should be mentioned also. The calculations show[(142,143)] that this effect is negligible if the particles carry not more than one elementary charge but can become significant in the case of larger charges. In their study of the diffusion method by means of a monodisperse aerosol obtained from polystyrene latex, Megaw and Wiffen[(144)] showed that correct results are obtained only with grounded metal batteries. Undoubtedly, latex particles can carry large charges. It was found in reference 40 that if the aerosol under investigation had a higher temperature than the battery walls, deposition at small μ was greater than under isothermal conditions, evidently due to thermophoresis. In working with very HDAs in batteries of different length it is important to keep the losses in communication lines constant. For this purpose, in reference 40 the change in the battery length was compensated for by moving the PSM from place to place by means of a trolley. The last measure was considered necessary since, according to the observations of the authors of reference 40 as well as of Nolan and Kuffel,[145] in bent rubber tubes particle deposition is much greater than in straight ones due to electric charges produced by the bending of the tubes.

Apart from Megaw and Wiffen, Thomas[135] also tested the diffusion method. He compared the results of optical and diffusional determinations of the particle size in the range of 1000–3000 Å and obtained satisfactory agreement. Some discrepancies observed can be due to incorrect compensation of the entrance effect. For larger particles, the diffusion method is inapplicable on account of increased sedimentation effect, which is very difficult to eliminate completely even by vertical arrangement of the batteries.

The diffusion method can be realized not only in channels, but in any system of bodies making possible a theoretical calculation of the convective diffusion deposition as, for example, in a model of fibrous filters consisting of a system of parallel cylindrical regularly spaced fibres oriented perpendicular to the flow. The penetration of HDAs through such a model is expressed by the formula

$$n/n_0 = \exp(-2\xi\eta H/\pi R_f), \tag{2.8}$$

where ξ is the volume fraction occupied by fibres, H is the model thickness, R_f is the fibre radius, and η is the capture coefficient of particles by fibres expressed by a theoretical formula.[146]

$$\eta = 2.9K^{-1/3}Pe^{-2/3}+0.624Pe^{-1}, \quad K = \tfrac{1}{2}\ln\xi+\xi-0.25\xi^2-0.75, \tag{2.9}$$

where $Pe = 2R_f\bar{v}/D$ and $\bar{v}$ is the mean flow velocity in the model. Equations (2.8) and (2.9) are in good agreement with experimental data[147] on penetration through model filters of monodisperse HDAs of $\bar{a}$ = 18–80 Å (Fig. 3). The diffusion coefficients of particles in these experiments were determined by means of diffusion batteries. The advantage of this version of diffusion method is the absence of entrance effects.

In reference 149 the diffusion method was combined with the determination of the number of charged particles, a diffusion channel being used as a plane-parallel condenser. The measurement of the percentage of particles charged under steady-state conditions determines the "equivalent" size of aerosol particles (see p. 21); a certain mean particle size can also be determined from diffusion measurements. The ratio of these two mean values can characterize qualitatively the polydispersity. The general idea of this method is true enough, but inasmuch as the values of D measured by the diffusion method depend on y, the result thus obtained will be dependent both on the aerosol parameters and on the conditions of measurements.

Some authors[150] used the so-called static diffusion method, in which the concentration change in time of a stationary aerosol contained in a cylindrical or spherical vessel is measured. For 10 Å particles this method is inapplicable, since such particles are deposited in the time of the order of a few seconds, whereas for larger particles it is less accurate than the dynamic method on account of errors due to convection.

Labeirie[151] suggested a method for investigation of aerosols, HDAs included, based on their radioactivation produced by mixing them with radon. After deposition of decay products on aerosol particles, the latter can be deposited on a filter and the number of particles counted by a radiographic method.

In conclusion, consider the methods of studying HDAs based on their preliminary deposition. The methods of investigating highly dispersed deposits do not depend on those by which they were obtained. Therefore there is no reason to consider here in detail such methods as, for example, electronic microscopy, since they are treated in extensive, special literature. We shall only discuss some specific features of examination of aerosol deposits. The main requirement here is that the sample chosen should be representative, i.e. the

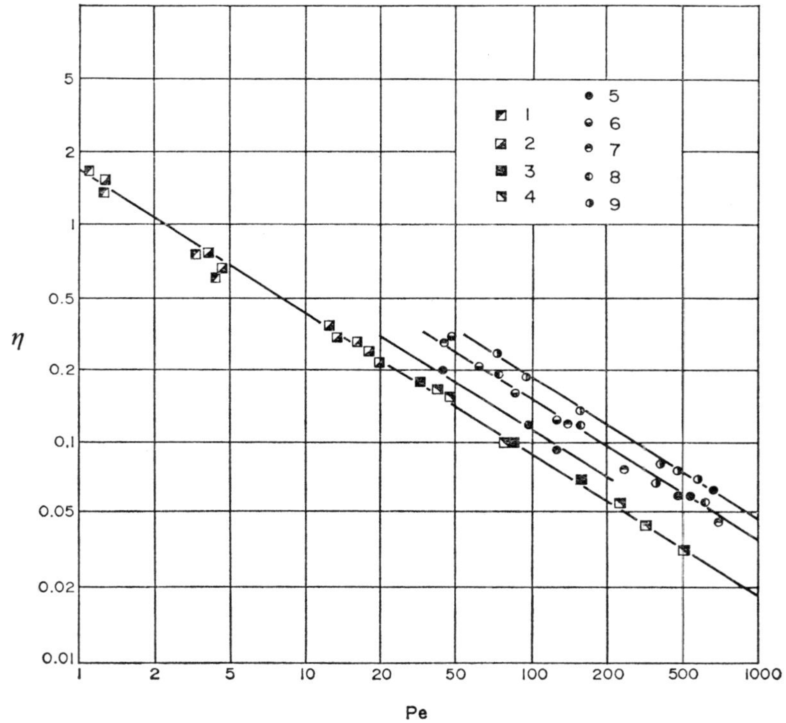

FIG. 3. Capture coefficient in the diffusional deposition of particles in a model filter with staggered arrangement of cylinders: theoretical curves. ξ: I 0.01, II 0.05, III 0.135, IV 0.27; mean particle radius $\bar{a}$, Å: 1, 15; 2, 18; 3, 60; 4, 60; 5, 70; 6, 41; 7, 70; 8, 55.

particle deposition should not be selective. In most cases, samples are taken by thermal precipitators, the design of such, ensuring representative samples, being described in references 152 and 153. The most representative samples are obtained in thermal precipitators with a moving wire.[154] Sometimes electric precipitators are used for deposition. Since in these instruments a certain segregation of particles takes place as aerosol moves along the depositing electrode, it is necessary to examine samples taken at various points of the electrode. For investigation of deposits of HDAs by means of the BET method, X-ray diffraction, etc., electric precipitation is used almost exclusively.

An ingenious design of an electrostatic precipitation classifier is described in reference 155. A thin jet of HDAs charged in a corona discharge is introduced into a laminar flow between plane-parallel or coaxial electrodes. The classification during deposition is due to the fact that the velocity of particle motion towards the depositing electrode is determined by the ratio of the charge and the drag coefficient of the particle, and they are not proportional to one another.

Electronic microscopy is an extremely important method of investigation of HDAs since it determines directly the size and shape of aerosol particles as well as the particle concentration. However, the use of electronic microscopy for the investigation of HDAs presents certain difficulties, which become greater with particle size decrease and are due to the melting and evaporation of particles in the electron beam as well as to poor contrast on account of greater transparency of particles. The well-known practices of electronic microscopy shadowing of particles by vacuum sublimation of heavy metals, shadowing of replicas—the sandwich method[156, 157]—give only a very limited effect since in investigation of very fine particles the thickness of the shadowing layer becomes commensurable with the size of particles. Therefore it is no wonder that although many authors studied the deposits of particles of the size corresponding to the limit of the resolving power of the electron microscope,[32, 44, 45] they prefer to illustrate their papers with photographs of particles of the size 100–1000 Å. In reference 47 it is pointed out that it is impossible to obtain sharp photographs of sodium chloride particles less than 200 Å. In reference 27 it is shown that platinum and nickel particles of less than 100 Å are either permeable to electrons or rapidly evaporate under their action. The attempts of the authors to examine, by means of an electron microscope with the resolving power limit 15 Å, the structure of MoO_3 aggregates consisting of individual particles of 100 Å (as was found by the BET method) proved to be unsuccessful. The problem of electronic microscopy of liquid particles should be considered as unsolved as yet.

In the cases when it is possible to collect a weighable quantity of aerosol deposit, the mean particle size can be calculated from the specific area of the deposit, determined by some conventional method. If during coagulation the particles did not sinter or coalesce but formed loose aggregates, it is possible to calculate from the deposit surface area the size of the primary, uncoagulated particles. The formation of compact aggregates, however, and, particularly, their sintering, can lead to too high values of the particles' size thus calculated. In references 65 and 66 the results obtained by the BET method were compared with those of an electronic microscopy, it being established that in the case of Al_2O_3, TiO_2, ZrO_2, Fe_2O_3, and SiO_2 deposits of aerosols obtained by interaction of the vapors of chlorides of these metals with water vapors at 1600–2000°C, the two methods gave similar results. The results of both methods are in good agreement[37] also for aerosol deposits of metal oxides obtained in an electric arc. On the other hand, a decrease in the surface area of the deposits of NaCl HDAs was reported in references 47 and 48 and of nickel in reference 38, which can be a serious source of errors, especially if considerable time passes between the formation of the deposit and its examination.

Membrane filters are also used in the investigation of HDAs.[158–162] Cartwright[162] proposed a method for preparation of replicas for electronic microscopy of the particles deposited on membrane filters. In the investigation of radioactive aerosols, radiography of filters is often used, membrane filters included. Frequently, membrane filters are used for detecting highly dispersed particles with ice-forming activity. For this purpose, the filter with particles deposited on it is immersed in a supercooled sugar solution, and ice crystals start to grow on the particles.

In conclusion it can be stated that the existing methods of investigation of HDAs ensure reliable determination of all parameters of HDAs consisting of solid particles.

CHAPTER 3

PROPERTIES OF HDAs

3.1. Transfer processes in HDAs: general considerations

Of most importance in the physics of aerosols are the processes of transfer from the medium to the particles of mass (evaporation, condensation growth, and Brownian coagulation of particles), heat (heating and cooling of particles due to heat conductivity of gaseous medium), momentum (resistance of medium to the movement of particles and the forces acting on particles in inhomogeneous medium), and charge (charging and discharge of particles by gas ions). The theory of these phenomena is simple in two extreme cases—at very large and very small Knudsen numbers. In the first case, the transfer processes are expressed by well-known equations of diffusion and heat conduction theory and hydrodynamics of continuous media. In the second case, transfer processes occur in the free molecular regime in which a particle does not affect the distribution function f of gas molecules moving towards it; for molecules moving in the reverse direction f is readily found from the law of reflection or evaporation of molecules from the particle surface.

The case of intermediate *Kn* values presents the greatest difficulties. For many years this problem was solved as follows: for transfer at a distance from a particle the equations of continuous media theory were used and, in the vicinity of the particle, the free molecular transfer equations. The mass, momentum fluxes, etc., calculated by means of these equations were conjugated either at the particle surface or at a distance of the order of the mean free path from it. Both versions of the method of flux conjugation gave qualitatively reasonable results, but in the light of the new studies it is evident that the method is inaccurate. However, for charge transfer only the conjugation method has been used so far.

It is possible to calculate accurately the transfer processes at intermediate *Kn* only on the basis of the solution of the Boltzmann kinetic equation. The chief difficulty in this case is known to be the collision integrals, where the products of distribution functions appear under the integral sign. For vapor diffusion through gas, in the simplest case when the ratio of vapor and gas molecule masses $m_v/m_g \to 0$ (see section 3.2), this integral degenerates to a simpler linear function of f. In the general case, in problems involving the solution of the kinetic equation, a method of its linearization has been used. This method, suggested by Welander[172] and by Bhatnagar, Gross, and Krook,[173] consists in the substitution for the collision integral of an expression which, in the simplest case, is of the form

$$I = (f_m - f)\nu/\beta, \tag{3.1}$$

where f_m is Maxwellian, f is the actual distribution functions, ν is the number of collisions between molecules in unit time and in unit volume, and β is a coefficient depending on the physical quantity transferred during collisions, the ratio of the masses of colliding molecules, etc. The physical significance of this simplification will be considered in the next section.

The linearization method has been further developed by other authors and successfully used for the solution of a large number of problems of theoretical physics.

Brock and Willis applied this method to the transfer processes for the case $Kn \gg 1$. They wrote the distribution function as

$$f = f_0 + f_1 Kn^{-1}, \tag{3.2}$$

where f is the distribution function in the free molecular regime. Substituting this expression into the linearized kinetic equation, equating the terms containing Kn^{-1}, and solving the differential equation thus obtained, Brock and Willis determined the correction factor by which the expression for the flux in the free molecular regime φ_k should be multiplied to pass to the flux at a finite Kn. The equations thus obtained are of a general form,

$$\varphi = \varphi_k(1 - \lambda Kn^{-1}) \approx \varphi_k(1 + \lambda Kn^{-1}), \tag{3.3}$$

and approximate well at $Kn \gg 1$ the general solutions obtained in a more complicated manner.

Recently, considerable progress has been made in the theory of transfer processes at intermediate Kn numbers. It should be noted that this progress is due to authors whose interests are far from HDAs. Sahni—who was engaged in research on the theory of transfer and absorption of neutrons in nuclear reactors—made a very accurate calculation of the rate of mass transfer to a sphere at any Kn on the basis of the so-called integral equation of this theory.

The transfer processes at large and intermediate Kn being of great interest for rocket engineering, this problem attracted the attention of a number of theoretical physicists. By means of the variational method developed by Cercignani and Pagani they succeeded in calculating accurately the heat flux to a spherical particle and the resistance of the medium to its motion at any Kn, the results of calculations in the second case being in very good agreement with experimental data.

Naturally, in the studies connected with rocket engineering, great attention is given to the transfer processes at large velocities of particles with respect to the medium and high temperature and concentration gradients. In this article we shall confine ourselves to the case of small values of these quantities and low velocities in particular as compared to the thermal speed of molecules.

In conclusion, it should be noted that in the theory of transfer processes to particles we have to use the notion of the mean free path of gas molecules, whereas in the modern theory of transfer processes in gases this notion is not used at all owing chiefly to its ambiguity. In essence, in the theory of transfer to particles l should be considered not as the mean distance passed by the molecules between consecutive collisions, but as the mean effective length of the free path l_{eff}, which can vary for different transfer phenomena. Thus, for mass transfer, l is the mean distance to which a molecule is transferred from the initial position at the moment when its velocity vector becomes independent of its initial value. In heat transfer, l_{eff} is the distance at which the molecule assumes the mean temperature of surrounding gas; in momentum transfer it is the distance at which the molecule assumes the mean (hydrodynamic) gas velocity. Without going into details, it can be stated, for example, that due to incomplete energy transfer corresponding to the internal degrees of freedom of molecules upon collisions, l_{eff} for heat transfer should be greater than for momentum transfer. So far, this important circumstance has been taken into consideration only by Lord and Harbour.[165]

3.2. Mass transfer in HDAs: evaporation and condensation growth of particles

Evaporation and condensation growth of particles plays a very important role in the formation and life of aerosols, and have been the object of numerous theoretical and experimental studies. In the following the problem is somewhat restricted, and the evaporation of a single spherical particle at rest in an unlimited gas volume is considered. This process can be considered as a quasi-stationary one if the relaxation time of vapor concentration around the particle is much less than the time of its complete evaporation, or, as shown by analysis, if the concentration of saturated vapor of the particle is much less than its density. We shall also assume that the vapor concentration is much less than that of gas, so that the collisions between vapor molecules and the Stefan flow from the particle surface can be neglected.

At a large enough distance from the particle, the vapor transfer occurs only by diffusion. The differential equation of steady vapor diffusion from a sphere or towards it is

$$\rho^2 \frac{dn}{d\rho} = I, \tag{3.4}$$

where n is the vapor concentration at the distance ρ from the center of the sphere and I is a constant determined from the boundary condition at the surface of the sphere.

Let us consider two simple cases—particle evaporation in a medium containing no vapor and condensation on a "black" sphere, i.e. a sphere absorbing all molecules colliding with it. Since in these cases n is proportional to the concentration n_s of saturated vapor of the particle and to the concentration n_∞ of vapor at $\rho = \infty$ respectively, by integrating eqn. (3.4), in the first case,

$$n = -\delta n_s/\rho, \tag{3.5}$$

and, in the second,

$$n = n_\infty(1-\delta/\rho), \tag{3.6}$$

where δ is a constant which is positive for condensation and negative for evaporation. The diffusion flow to the sphere (and from it) is expressed by the formulas

$$\varphi = 4\pi D\delta n_\infty \quad \text{and} \quad \varphi = 4\pi D\delta n_s. \tag{3.7}$$

If in the case of evaporation vapor concentration at infinity is equal to n_∞ rather than to 0, in all relevant formulas $(n-n_\infty)$ should be substituted for n, and (n_s-n_∞) for n_s. A similar substitution should be made for the case of condensation on a particle with the vapor pressure n_s rather than 0.

All these equations are valid at any Kn, but only at a large distance from the sphere. However, in the case $Kn \to 0$, vapor transfer is of a purely diffusional nature up to the sphere surface. The boundary condition for evaporation is $n_{\rho=a} = n_s$, and for condensation on a black sphere $n_{\rho=a} = 0$. Thus, in this case, $\delta = \pm a$ and eqn. (3.7) turns to the well-known Maxwell equations

$$\varphi_c = 4\pi Dan_\infty \quad \text{or} \quad \varphi_c = 4\pi Dan_s. \tag{3.8}$$

It should be noted that at $Kn \to 0$ the mass transfer rate does not depend on the coefficient of evaporation (condensation) α_c. However, according to the recently published paper of Jer Ru Maa,[167] α_c is close to unity for all liquids.

In the case $Kn \to \infty$, the rate of evaporation of a particle (at $n_\infty = 0$) and the rate of condensation on it (at $n_s = 0$) is expressed by Knudsen equation for evaporation under vacuum

$$\varphi_k = \pi a^2 \alpha_c \overline{V}_v n_s \quad \text{or} \quad \varphi_k = \pi a^2 \alpha_c \overline{V}_v n_\infty, \tag{3.9}$$

where $\overline{V}_v$ is the mean velocity of thermal motion of vapor molecules. The subscripts c and k stand for transfer in continuous and in free molecular regime respectively. Note that

$$\varphi_c/\varphi_k = \tfrac{4}{3}Kn\alpha_c^{-1}. \tag{3.10}$$

As already pointed out, the first attempts to solve the problem of particle evaporation at intermediate Kn consisted in conjugating the free molecular and diffusional fluxes. In a number of studies(168–170) the fluxes were conjugated directly at the particle surface, i.e. the diffusion equation was assumed to be valid up to the very surface. The formula thus obtained,

$$\varphi = \frac{\varphi_c}{1+\frac{4}{3}Kn/\alpha_c} = \frac{\varphi_k\alpha_c}{1+\frac{3}{4}Kn^{-1}\alpha_c}, \tag{3.11}$$

gives a correct functional dependence of φ on Kn but too high values of the coefficients before Kn and Kn^{-1}.

In the boundary sphere method(166) conjugation was made at the surface of the sphere surrounding the particle and located at the distance βl from the particle surface to which, on the average, the evaporating vapor molecules that have passed the distance are removed from the particle (at $Kn \to 0$, $\beta = \frac{1}{2}$; at $Kn \to \infty$, $\beta = 1$). In this case

$$\varphi = \frac{\varphi_c}{(4Kn/3\alpha_c)+1/(1+\beta Kn)}.$$

At $Kn \ll 1$ and $\alpha_c = 1$,

$$\varphi = \frac{\varphi_c}{1+\frac{2}{3}Kn}, \tag{3.12}$$

which is very close to the accurate expression. However, at $Kn \gg 1$ an incorrect functional dependence is obtained:

$$\varphi = \frac{\varphi_k}{1+\frac{3}{4}Kn^{-2}}. \tag{3.13}$$

Thus we see that the method of conjugation does not give correct results. Let us note also that the "universal" Scherman's interpolation formula(171) suggested by him for a number of transfer processes,

$$\varphi = \frac{\varphi_c}{1+\varphi_c/\varphi_k} = \frac{\varphi_k}{1+\varphi_k/\varphi_c}, \tag{3.14}$$

at $\alpha_c = 1$ coincides with eqn. (3.10).

A rigorous solution of the problem of particle evaporation at intermediate Kn is possible only on the basis of the Boltzmann kinetic equation. Under the condition specified above that vapor concentration is small compared to that of gas, the distribution function of gas molecules is only slightly perturbed by the collisions with vapor molecules and can be considered to a good approximation to be Maxwellian up to the particle surface. For vapor molecules, however, the Maxwellian distribution is valid only at a large distance from the particle surface, as compared to l. In the case of a "black" sphere, at its very surface the component of the vapor molecules velocities normal to the surface is directed towards it, i.e. here the distribution function is described by one "half" of the Maxwellian distribution directed towards the sphere. As the distance from the sphere increases, this asymmetry of the distribution gradually disappears. Vapor transfer at a large distance from the sphere is purely diffusional, i.e. determined by the vapor concentration gradient. As the distance to the sphere decreases, the diffusion flux of vapor is superimposed by the flux due to the asymmetry of the distribution function, this latter flux alone remaining at the sphere surface.

In the problem under consideration, the method of linearization of the kinetic equation mentioned in section 3.1 acquires a simple physical significance. Instead of calculating in detail the changes in the distribution function of vapor molecules caused by their collisions with gas molecules (according to what has been said previously the collisions between vapor molecules can be neglected), we assume that in these collisions vapor molecules assume the Maxwellian (isotropic) velocity distribution obeyed by gas molecules. It should be emphasized that this assumption can be true only in the case if the ratio of the masses of vapor and gas molecules $m_v/m_g \to 0$. In the opposite case, a certain persistence of vapor molecule velocities exists, and their velocity distribution after collisions cannot be completely isotropic.

Brock[174, 175] applied this method of linearization of the kinetic equation in conjunction with the Knudsen iteration to the theory of mass transfer at $Kn \geqslant 1$. Brock gives only the final formula without computations. The correction term of this formula contains the mean free path of gas molecules l_g. Since all the formulas in this section contain the mean free path l_v of vapor molecules, we have transformed Brock's formula by means of the well-known gas-kinetic equations

$$l_g = (\sqrt{2}\pi n_g d_g^2)^{-1}, \tag{3.15}$$

$$l_v = [\pi n_g d_{vg}^2 (1 + m_v/m_g)^{1/2}]^{-1}, \tag{3.16}$$

where subscripts v and g denote vapor and gas, d is the molecule diameter, and $d_{vg} = 0.5(d_v + d_g)$. In this case, Brock's formula assumes the following form:

$$\varphi = \varphi_k[1 - \alpha_c 0.807\sqrt{(2)}\theta Kn^{-1}], \tag{3.17}$$

where θ is a factor associated with the persistence of vapor molecules, equal to 0.295, 0.150, and 0.115 at $m_v/m_g = 0$, 1, and 9 respectively. As is clear from the above and as Brock himself points out, these calculations, strictly speaking, are applicable only in the case $m_v/m_g \to 0$ when eqn. (3.17) can be written as (at $\alpha_c = 1$):

$$\varphi = \varphi_k(1 - 0.42 Kn^{-1}) \approx \varphi_k/(1 + 0.42 Kn^{-1}). \tag{3.18}$$

As we shall see below, this formula at $Kn \gg 1$ gives a more or less correct value of φ.

From elementary considerations it is possible to find the lower limit for the coefficient before Kn^{-1}. Let us calculate the probability of the event that the evaporated molecule will be emitted to the distance l along the normal to the particle surface, will collide here with a gas molecule, and after isotropic scattering will return back to the particle without undergoing new collisions. Evidently, this probability is proportional to the ratio of the solid angle, at which the particle is visible at the distance indicated, to 4π. Taking into consideration the mean free paths' distribution of molecules on their way from the particle and back, we shall obtain for the mean probability of return of the molecules the expression

$$\overline{W}_1 = \frac{1}{2l_v}\int_0^\infty \left\{1 - \left[\sqrt{\left(\frac{x^2}{a^2} + 2\frac{x}{a}\right)}\Big/ 2\left(1 + \frac{x}{a}\right)\right]\right\} e^{-2x/l_v}\, dx. \tag{3.19}$$

At small Kn^{-1}, $\overline{W}_1 = 0.285\, Kn^{-1}$. If we had taken into account the fact that molecules evaporate from the surface at various angles rather than normal to it, we should have obtained a somewhat higher value for $\overline{W}_1$. Evidently, the total probability of return of molecule to particle W is greater than W_1. Since the rate of particle evaporation with allowance for returns is $\varphi_k = 1 - W$ the true value of the coefficient before Kn^{-1} in any case should be evidently greater than 0.285.

When the scattering of vapor molecules is non-isotropic and their velocities show persistence upon collisions, the probability of their return to the particle after the first collision evidently decreases, this decrease being the greater the stronger is the persistence, i.e. the greater is the ratio m_v/m_g. Thus, in eqn. (3.19) l_v should mean the "effective" mean free path of molecules l_{eff} (see section 3.1). Apparently, eqn. (3.18) should be treated in a similar way until an accurate transfer theory with account taken of the velocity persistence has been developed.

As was shown by Jeans,[176]

$$l_{\text{eff}} = l_v \frac{m_v(m_g+1)}{1+\beta}, \tag{3.20}$$

where l_v is the mean distance traveled by vapor molecules between two collisions, β is a number equal to 0 at $m_v/m_g \to 0$, increasing with m_v/m_g and reaching the value 1/3 at $m_v/m_g \to \infty$. It follows from eqns. (3.20) and (3.16) that

$$l_{\text{eff}} = \frac{\sqrt{(m_v/m_g+1)}}{(1+\beta)\,\pi n_g d_{vg}^2}. \tag{3.21}$$

By substituting in Meyer's formula for the vapor diffusion coefficient in gas,

$$D = \bar{V}_v l_v/3, \tag{3.22}$$

l_v by l_{eff}, Jeans' formula[176] is obtained,

$$D_v = \frac{\bar{V}_v\sqrt{(1+m_v/m_g)}}{3(1+\beta)\pi n_g d_{vg}^2}, \tag{3.23}$$

which differs from the well-known Stefan–Maxwell equation only by a factor $(1+\beta)$ close to unity. Thus it is best to use in eqn. (3.18) the value of l_v calculated from the coefficient of mutual vapor–gas diffusion by means of eqn. (3.22).

A similar problem to that under consideration, though a simpler one, was solved in the theory of neutron transfer. In Milne's spherical problem a neutron flux to a "black" sphere, i.e. a sphere absorbing all neutrons which come into contact with it, is considered. All neutrons are assumed to have the same absolute velocity and are isotropically scattered by heavy stationary atoms filling the space around the sphere. In this case, the velocity distribution function contains only the velocity directions. As the result of collisions, the distribution becomes independent of direction, and the Maxwellian distribution function assumes the form $\frac{1}{4}\pi$. Therefore the collision integral changes to the expression $(\frac{1}{4}\pi - f)\nu$, where ν is the number of collisions of neutrons with atoms in unit volume per unit time.[177]

Milne's spherical problem was solved in a number of studies. The most accurate and complete solution is that of Sahni,[178] who proceeded from the so-called "integral" equation of the theory of neutron transfer and used an analytical method of solution in conjunction with a computational one.

It follows from Sahni's calculations that at the distance from the sphere surface two to three times as large as l, the velocity distribution is practically isotropic and neutron transfer occurs exclusively by diffusion due to the concentration gradient. As the distance to the sphere decreases, the velocity distribution becomes more and more anisotropic, and diffusion is superimposed by directed flux: quite close to the sphere all neutrons move towards it, and transfer is effected entirely by directed flux.

The result of Sahni's work can be given by the formula

$$\varphi = \frac{4\pi a D n_\infty}{1+\lambda Kn}. \tag{3.24}$$

The values of λ calculated by Sahni are listed in Table 1. According to the author, they are accurate within $\frac{1}{3}\%$.

TABLE 1. VALUES OF THE COEFFICIENT λ ACCORDING TO SAHNI

Kn^{-1}	0	0.1	0.2	0.5	0.7	1.0	1.3	1.5	2.0	2.5	5.0	∞
λ	1.333	1.296	1.253	1.155	1.104	1.043	0.997	0.972	0.925	0.892	0.813	0.710

For $Kn \ll 1$, from eqn. (3.24)

$$\varphi = \frac{\varphi_c}{1+0.710Kn}. \tag{3.25}$$

In the case $Kn \gg 1$ it is necessary to express λ according to Table 1 as $\lambda = 1.333 - 0.37\,Kn^{-1}$ and to substitute $\overline{V}_v l_v/3$ for D. The result is

$$\varphi = \frac{\varphi_k}{1+0.472Kn^{-1}}, \tag{3.26}$$

which is similar to Brock's formula.

Since in calculations of the evaporation kinetics it is inconvenient to use Table 1, it can be substituted by an interpolation formula

$$l = \frac{1.333+0.71Kn^{-1}}{1+Kn^{-1}}. \tag{3.27}$$

The maximal deviation of the values calculated by means of this formula from those given in Table 1 is 2–6%. In calculating the particle evaporation time by means of eqn. (3.27), we obtain an easily taken integral of a rational function of a.

Since the mean absolute velocity of molecules $\overline{V}_v$ does not figure in the calculation of the value of λ, the restrictions imposed by us above, the assumption of equal absolute velocity of all molecules is irrelevant: the velocity distribution can be of any type, the Maxwellian included. It is necessary only to preserve during collisions the isotropy of scattering, i.e. the condition $m_v/m_g \to 0$. Since l depends on $\overline{V}_v$, an averaged value of l should be used in eqn. (3.24).

In the theory of neutron transfer the case of their velocity persistence is also treated.[177] However, Milne's spherical problem taking account of persistence has not been solved as yet and, for lack of a better method, persistence is to be allowed for as described above on p. 34.

The condensation coefficient α_c being for some solid substances less than 1, it is necessary to consider also the case of a "gray" sphere absorbing only a fraction α_c of the vapor molecules colliding with it. It is easily seen that the gray sphere is equivalent to a sphere having a certain vapor pressure and $\alpha_c = 1$. Let us denote by n_- and n_+ the concentrations of vapor molecules at the gray sphere surface, moving towards and from it respectively. Evidently,

$$n_+ = n_-(1-\alpha_c). \tag{3.28}$$

The flux of molecules towards and from the sphere can be expressed by the formulas

$$\varphi_+ = 2\pi a^2 \overline{V}_v n_- = 2\pi a^2 \overline{V}_v n_+ + 2\pi a^2 \overline{V}_v(n_- - n_+) \tag{3.29}$$

$$\varphi_- = 2\pi a^2 \overline{V}_v n_+. \tag{3.30}$$

But we shall have the same fluxes in the case of a sphere with vapor pressure $2n_+$ at $\alpha_c = 1$ and with the resulting flux towards the sphere equal to

$$\varphi_{gr} = 2\pi a^2 \overline{V}_v(n_- - n_+). \tag{3.31}$$

Thus these two spheres are equivalent, i.e. the vapor concentration distribution about them are identical. As it follows from what was said on p. 34, the resulting flux may be written also as

$$\varphi'_{gr} = \frac{4\pi a D(n_\infty - 2n_+)}{1 + \lambda l/a}. \tag{3.32}$$

Eliminating n_+ from eqns. (3.28) and (3.31) and equating φ'_{gr} to φ_{gr}, it is possible to find an expression for n_+, and hence the formula for the flux towards the gray sphere

$$\varphi_{gr} = \frac{\varphi_D}{1 + \left[\lambda + \dfrac{4(1-\alpha_c)}{3\alpha_c}\right] Kn} \tag{3.33}$$

obtained by Smirnov by a different method.[179]

We shall return to the problem of the effect of capillary forces on the evaporation of fine particles in section 3.8.

Before passing to the comparison of the theory outlined here with scarce experimental data available, it should be noted that direct measurements of the evaporation rate of HDA particles present very great difficulties and thus call for low pressures of gaseous medium (to increase l_v). The well-known experiments of Bradley and collaborators[180] and of Monchik and Reiss[181] cannot be considered inconsistent with the theory outlined above, but their accuracy is too low for solving the yet unsolved problem of the effect of the persistence of vapor molecules velocity on the mass transfer rate.

The calculation of the growth or evaporation of aerosol particles under actual conditions involves taking into account the heat release of the phase transition, the effect of the particles' motion, and other factors. These problems have been considered by a number of authors. Brock calculated the condensation flux to a particle moving in a free molecular regime[182] and took account of the condensation heat release in free molecular growth.[175] By means of the Welander–Bhatnagar's model, he studied the non-steady-state particle growth at intermediate Kn numbers. Smirnov[184] obtained expressions for the growth or evaporation rates of particles allowing for the phase transition heat and some minor factors. Kang Sang Wook[185] used the boundary sphere method for calculation not only of the particle growth rate, but also of the dissipation rate of condensation heat. The results of calculations by means of his formula were compared with experiments on the mercury vapor condensation in a nozzle.[186] This comparison, however, does not prove the correctness of the formula obtained in reference 185.

Mention should be made also of the method of calculating the evaporation rate of a system of aerosol particles suggested by Okuyama and Zung[187] and of the rate of free molecular evaporation of water drops injected into a highly rarefied plasma.[188]

In conclusion, let us discuss briefly the adsorption of gases and vapors on HDA particles. Evidently, the equations for vapor condensation on particles can also be used to describe the adsorption kinetics. The only difference should be the dependence of the sticking probability on the amount of sorbed substance, i.e. on the surface coverage with adsorbate. The total process of vapor adsorption on aerosol particles can be calculated provided the desorption kinetics is known.

In reference 189 the boundary sphere method was used to estimate the adsorption rate of iodine vapors by atmospheric condensation nuclei. It was found that iodine should be sorbed practically completely in 30 sec. In practice, however, owing to simultaneous desorption, the iodine fraction sorbed by particles was no more than 75% of total iodine in the atmosphere, this value being attained only after $2\frac{1}{2}$–3 hr.

As to the decay products of emanations, they are practically completely ionized,[190] so that the process of their capture should be described by the equations of the charging rate of HDA particles by gas ions (see section 3.5).

3.3. Momentum transfer in HDAs: resistance of the medium to the motion of particles

In the theory of transfer in aerosols this process stands apart since it is the only case for which accurate and quite reliable experimental data obtained by Millikan[191] for oil droplets are available. This makes possible an accurate verification of theoretical computations.

At low enough Reynolds numbers the resistance of the medium at $Kn \to 0$ is expressed by Stokes' formula

$$F_c = 6\pi\eta a U, \tag{3.34}$$

where η is the gas viscosity and U is the particle velocity.

For $Kn \to \infty$, Epstein[192] derived the formula

$$F_k = \frac{4}{3}\pi\beta n_g m_g a^2 \bar{v}_g V, \tag{3.35}$$

where β is a coefficient whose value depends on the mechanism of reflection of molecules from the particle. With a complete thermal accommodation and diffusive scattering of molecules, $\beta = 1+\pi/8$, with a specular reflection $\beta = 1$. Substituting into eqn. (3.35) the well-known expression for gas viscosity $\eta = 0.499 n_g m_g \bar{V}_g l_g$, eqn. (3.35) is written as

$$F_k = 6\pi a^2 \eta U/(A+Q) l_g, \tag{3.36}$$

where the quantity $(A+Q)$ has for the two cases mentioned the values 1.616 and 2.250 respectively. It follows from eqns. (3.34) and (3.36) that

$$F_c/F_k = (A+Q)Kn. \tag{3.37}$$

Millikan expressed the results of his experiments with oil droplets by means of an empirical equation

$$F = F_c = [1+AKn+QKn \exp(-bKn^{-1})]^{-1} \tag{3.38}$$

in which $A = 1.234$, $Q = 0.414$, and $b = 0.876$. Obviously, at $Kn \to \infty$ eqn. (3.38) turns to eqn. (3.36). From the comparison of the theoretical values of $A+Q$ with the experimental value 1.65, it is possible to draw a formal conclusion that a small fraction of molecules is reflected specularly from the surface of the particles, whereas the rest undergo complete accommodation and diffusive scattering.

The first attempt to calculate theoretically the resistance of the medium at intermediate Kn numbers was made by Fuchs and Stechkina,[193] who used the boundary sphere method. On the basis of some not quite vigorous assumptions, they obtained for the resistance to the motion of oil droplets the formula

$$F/F_c = \left[\frac{1}{1+0.42Kn} + 1.67Kn\right]^{-1}. \tag{3.39}$$

The values of F/F_c calculated by means of this formula differ from Millikan's data by no more than 2%.

Expanding the collision integral in the kinetic equation into Hermite polynomials and using Knudsen iteration, Liu and collaborators[194] derived the formula

$$F/F_k = 1-0.298Kn^{-1}. \tag{3.40}$$

By linearizing the kinetic equation, as indicated in section 3.2, Willis[195] obtained by means of Knudsen iteration the expression

$$F/F_k = 1-0.366Kn^{-1}. \tag{3.41}$$

Cercignani and Pagani[196] use a new method for the solution of transfer problems at intermediate *Kn*. After linearization of the kinetic equation they build a functional of the functions expressing gas density, temperature, and velocities of molecules and find the stationary value of the functional by the variational method. Then the physical quantities assume the values corresponding to the solution of the linearized equation, and the fluxes of momentum, heat, etc., are calculated directly from the stationary value of the functional. By means of this method, the above-mentioned authors solved successfully a number of transfer problems in systems of different geometry.

Recently Cercignani and Pagani[197] applied their method to the solution of the problem of the resistance of the medium to the motion of spheroidal particles. The results of their calculations for spherical particles are shown in Fig. 4 as the dependence of F/F_k on Kn^{-1}. The calculations approximate within 2% the experimental formula (3.38). Undoubtedly,

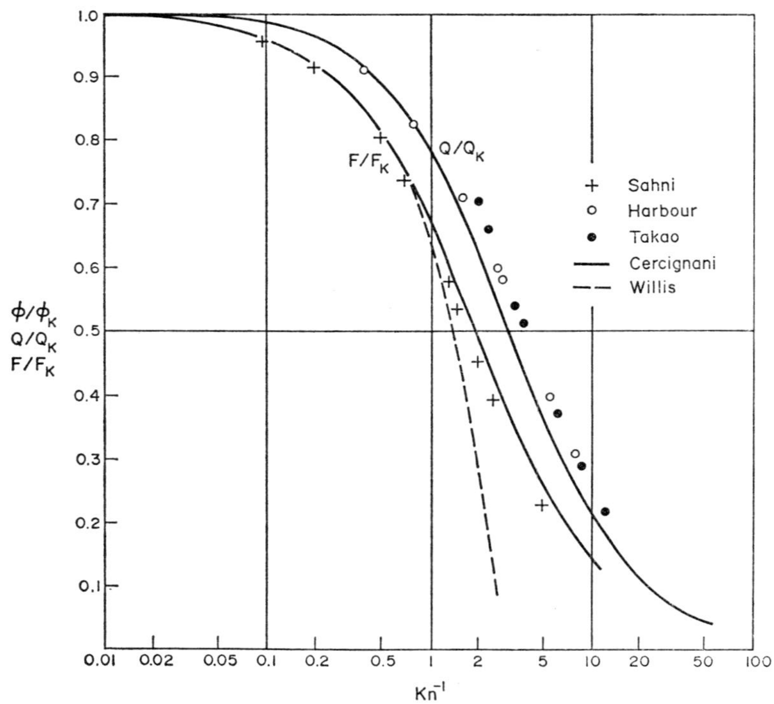

FIG. 4. Kinetics of transfer processes at intermediate *Kn*.

the work of Cercignani and Pagani has made a very important contribution to the theory of transfer phenomena at intermediate Kn.

It is evident from Fig. 4 that Willis's formula [eqn. (3.41)] gives (as would be expected) correct values of F/F_k only up to $Kn^{-1} \approx 0.5$, whereas eqn. (3.40) is obviously incorrect. As regards Sherman's "universal" formula [see eqn. (3.14)]

$$F/F_k = 1(1+F_k/F_c), \tag{3.42}$$

which can be written, using eqn. (3.37), as

$$F/F_k = [1+Kn^{-1}(A+Q)^{-1}]^{-1} = [1+0.606\ Kn^{-1}]^{-1},$$

it gives somewhat too low (up to 7%) values of F/F_k. It is interesting that the values of φ/φ_k as a function of Kn^{-1} calculated from Sahni's data for mass transfer (from Table 1) by means of eqn. (3.10) fall quite well on the curve for momentum transfer (Fig. 4). At the same time, the curve for heat transfer lies much higher.

Brock[(198)] obtained a very complicated expression for the resistance of the medium to the motion of a growing or evaporating particle at large Kn, which undoubtedly will prove helpful in calculation of aerosol formation in rocket nozzles and of the effect of the particles formed on the flow properties.

3.4. Heat transfer in HDAs: heating and cooling of particles

The process of heat transfer from the surrounding medium is primarily important in that the temperature of an evaporating particle, or of a particle growing due to condensations, differs from the temperature of the medium, and in the case of substances with appreciable vapor pressure this circumstance affects considerably the kinetics of evaporation and growth of particles.

At $Kn \to 0$ the rate of heat transfer to a particle is expressed by a formula derived in the classical theory of heat conductivity, which is similar to eqn. (3.8),

$$Q_c = 4\pi a f \Delta T, \tag{3.43}$$

where f is the thermal conductivity coefficient and ΔT is the temperature difference between medium and particle. We shall confine ourselves to the consideration of the simplest case, when ΔT is small compared to the absolute temperature of the particle T.

To derive the expression for the heat flux to the particle at $Kn \to \infty$, let us introduce the superscripts + and − for molecules moving from and towards the particle respectively. It follows from the condition of material balance that $n_g^+ V_g^+ = n_g^- v_g^-$ or

$$n_g^+ \sqrt{T^+} = n_g^- \sqrt{T^-}. \tag{3.44}$$

The flux of molecules to the particle is equal to

$$\pi a^2 n_g^- \bar{V}_g^- = \pi a^2 n_g^- (8kT^-/\pi m_g)^{1/2}$$

and the mean thermal energy of a molecule $c_v T^-$, where $c_v = 3/2k$ for monatomic and $5/2k$ for diatomic gases. The energy flux to a particle cannot be found, however, by multiplying these two quantities, since the greater is the kinetic energy of molecules, the relatively more frequent are their collisions with the particle. An accurate expression for the energy flux[(199)] is of the form

$$I^- = \pi a^2 n_g^- \left(\frac{8kT^-}{\pi m_g}\right)^{1/2} \left(c_v + \frac{k}{2}\right) T^-. \tag{3.45}$$

A similar expression (with superscripts + substituted for −) is obtained also for the energy flux carried away by molecules, the resulting flux being equal to the difference of these two expressions,

$$Q_k = \pi a^2 \left(\frac{8k}{\pi m_g}\right)^{1/2} \left(c_v + \frac{k}{2}\right) [n_g^-(T^-)^{3/2} - n_g^+(T^+)^{3/2}]. \tag{3.46}$$

Let us introduce the temperature accommodation coefficient of molecules $\alpha_T = \Delta'T/\Delta T$, where $\Delta'T$ is the difference between the mean temperature of the molecules leaving the particle and that of the medium. Evidently, $\Delta'T = T^- - T^+$. Assuming $\Delta'T/T$ to be a small quantity, we obtain from eqn. (3.46) by means of condition (3.44) the formula

$$Q_k = \pi a^2 \left(\frac{8kT}{\pi m_g}\right)^{1/2} \left(c_v + \frac{k}{2}\right) n_g \Delta'T = \pi a^2 \alpha_T \left(c_v + \frac{k}{2}\right) \Delta T n_g \overline{V}_g. \tag{3.47}$$

Let us find the value of the ratio $\alpha_T = \Delta'T/\Delta T$. From the well-known formula for heat conductivity of gases, $f = \varepsilon \eta c_v/m_g$, where η is the gas viscosity and the coefficient ε is 2.5 for monatomic and 1.9 for diatomic gases, and from the formula

$$\eta = 0.5 m_g n_g \overline{V}_g l_g$$

$$f = 0.5 \varepsilon c_v n_g \overline{V}_g l_g, \tag{3.48}$$

whence it follows that

$$Q_c/Q_k = \frac{2\varepsilon c_v Kn}{\left(c_v + \frac{k}{2}\right)\alpha_T}. \tag{3.49}$$

Thus for monatomic gases $Q_c/Q_k = 3.75 Kn \alpha_T^{-1}$ and for diatomic $Q_c/Q_k = 3.17 Kn \alpha_T^{-1}$.

For intermediate Kn, Springer and Tsai[200] obtained by means of the boundary sphere method the formula

$$Q/Q_c = \left(\frac{1}{1+Kn} + BKn\alpha_T^{-1}\right)^{-1}, \tag{3.50}$$

where $B = 3.75$ for monatomic and 3.17 for diatomic gases. At $Kn \to \infty$ this formula assumes the form $Q/Q_c = (BKn\alpha_T^{-1})^{-1}$. Taking into consideration the expressions for Q_v/Q_k we have for this case $Q/Q_k \to 1$, as it should be.

Using the momentum method, Lees[201] obtained for diatomic gases the formula

$$Q/Q_c = (1 + 2.5 Kn)^{-1}, \tag{3.51}$$

which is obviously inapplicable at large Kn.

Sherman's "universal" formula in this case is of the form

$$Q/Q_k = \frac{1}{1 + Q_k/Q_c} = \frac{1}{1 + \beta\alpha_T Kn^{-1}}. \tag{3.52}$$

For the value of β see eqn. (3.49).]

Using the Knudsen iteration method, mentioned earlier, Brock[202] deduced the formula

$$Q/Q_k = 1 - \alpha_T BKn^{-1}, \tag{3.53}$$

where $B = 0.153$ for monatomic and 0.239 for diatomic gases. As we have already pointed out, such formulas are accurate only at large Kn.

Using the variational method mentioned in the previous section, Cercignani and collaborators[203] calculated (assuming $\alpha_T = 1$) the values of Q/Q_k for monatomic gases in the range of Kn from 0.0001 to 50. Figure 4 shows the curve plotted from these data.

Until quite recently experimental determination of the heat transfer rate towards a sphere at small Re and intermediate Kn has been performed only by Takao.[204] He measured the cooling rate of a heated glass sphere 1.5 cm in diameter, placed in a vessel approximately 1 m^3 in volume, from which air had been evacuated until the desired residual pressure was reached. All the authors mentioned above compared their formulas (Takao also derived a very complicated formula) with Takao's data and all found a very good agreement. Takao's data seem to be accurate enough only at small Kn and incorrect at $Kn \gg 1$, which is evidenced by a large scatter of experimental points in this range. For $Kn > 1$ Harbour's[205] recently published results (also for air) seem to be more accurate. Unfortunately, the values of α_T in Takao's and Harbour's experiments are unknown, which makes comparison of theory with experiment impossible. In addition to Cercignani's curve, Fig. 5 shows Brock's curve (3.53) almost coinciding with it at large Kn, as well as experimental data of Harbour and Takao.

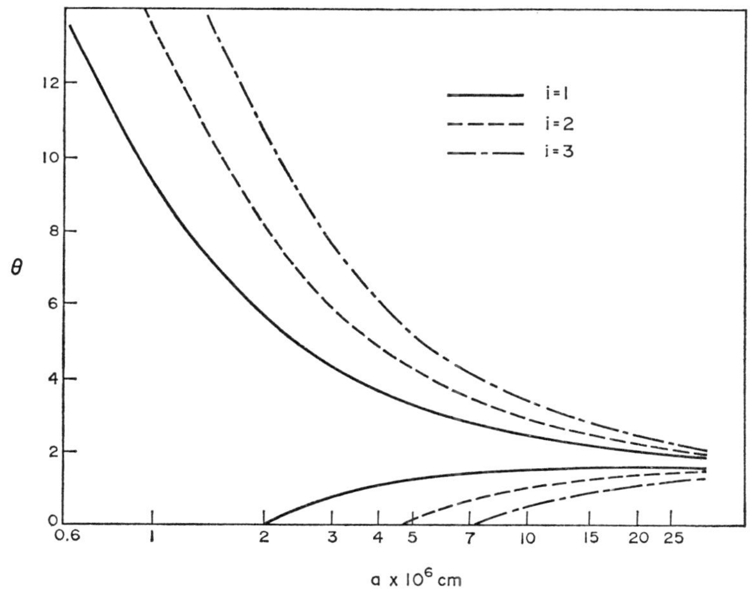

FIG. 5. Electrostatic factor in the ion diffusion to charged particle.

As pointed out in section 3.1, in calculating heat transfer it is necessary to use the value of the mean free path of molecules specifically for this process, l_H, rather than the generally accepted value for momentum transfer l_m. According to Lord and Harbour,[165] for diatomic molecules $l_H/l_m \approx 1.6$. If, however, Cercignani's curve should be referred to l_m rather than to l_H, it should be shifted to the left so as to decrease the abscissa 1.6 times. Then this curve will nearly coincide with respective curves for mass and momentum transfer, and we shall have practically one curve for all the three processes.

3.5. Charge transfer in HDAs: charging and discharging of particles by gas ions

One of the most important physical processes in aerosols is the charge transfer to particles by gas ions captured by them. It should be noted that we do not exactly know as yet what a gas ion actually is. There is considerable evidence that primary ions, formed upon emission or adsorption of an electron by gas molecules, add, under the action of electric forces, neutral molecules possessing a constant dipole moment, thus forming clusters. Electric mobilities of atmospheric "light" ions range within 0.5–2.5 cm^2/v sec. Hence it follows that the number of molecules in ion clusters also varies considerably.

Since the expressions for electric mobility and diffusion coefficient of ions[199]

$$B = \tfrac{3}{8}\pi^{\frac{1}{2}} l_i \varepsilon \left(\frac{m_g+m_i}{m_g m_i kT}\right)^{\frac{1}{2}}, \tag{3.54}$$

$$D = \tfrac{3}{8}\pi^{\frac{1}{2}} l_i \left[\frac{(m_g+m_i)kT}{m_g m_i}\right]^{\frac{1}{2}}, \tag{3.55}$$

contain both the mean free path of ions l_i and their mass m_i (m_g is the mass of neutral gas molecules), we can find l_i only by choosing a definite value for m_i. If we assume $m_i = m_g$ and take for ion mobility the mean value $B = 1.4\ cm^2/v$ sec, we shall have in air at 20° and normal pressure $l_i = 1.3\times 10^{-6}$ cm, and for mean thermal velocity $\bar{V}_i = 4.62\times 10^4$ cm/sec.

On the other hand, if we assume, as Junge does,[206] that $m_i/m_g = 6$–18, we shall have for l_i the value 1.8×10^{-6} cm, for $\bar{V}_i$ 1.0–2.0 $\times 10^4$ cm/sec. This uncertainty in the values of $\bar{V}_i$ and l_i makes comparison with experimental data extremely difficult. It should be also noted that the regime of the charge transfer to particle is determined by the "ionic Knudsen number" Kn, which is several times less than the "molecular" Kn, so that the upper limit of HDA particle radius, 10^{-5} cm chosen by us, corresponds to $Kn_i = 0.13$–0.18.

Let us consider first the charging of particles by ions, all of the same sign, in the absence of external electric field. The differential equation of the steady-state diffusion of ions towards the particle in the electric field produced by it is of the form

$$\varphi = 4\pi\rho^2\left(D\frac{dn}{d\rho} - BEn\right) = \text{const} \tag{3.56}$$

where E is the field strength. At $Kn_i = 0$ we have the boundary conditions: $n = n_\infty$ at $\rho \to \infty$ and $n = 0$ at $\rho = a$, and arrive at the formula

$$\varphi_c = \frac{4\pi D n_\infty}{\int\limits_a^\infty \frac{1}{\rho^2}\exp\left[\frac{\varphi(\rho)}{kT}\right] d\rho}, \tag{3.57}$$

where

$$\varphi(\rho) = \frac{i\varepsilon^2}{\rho} - \frac{\varepsilon^2 a^2}{2\rho^2(\rho^2 - a^2)} \tag{3.58}$$

is the potential of the electric forces acting upon the ion, ε is an elementary charge, i is the number of elementary charges on a particle, and ρ is positive when ion and particle have like charges and negative when the charges are opposite. The first term in eqn. (3.58) corresponds to the Coulomb force and the second to the image force. The particle is being assumed to be an electric conductor, which for very small particles is justified in most cases.

In the deduction of eqn. (3.57) it is assumed that ion diffusion is a quasi-steady process. Below we shall see under what conditions this assumption is valid.

In order to solve the question whether it is possible to neglect the image forces in the theory of charging of HDA particles, let us calculate their effect for the case of ion diffusion to an uncharged particle ($i = 0$). The values of the integral in eqn. (3.57) in this case are 0.97, 0.83, and 0.62 for $a = 10^{-4}$, 10^{-5}, and 10^{-6} cm respectively. Hence it is clear that the image forces should not be neglected.

Passing to the case $Kn_i \to \infty$, let us note that in charging of particles by gas ions the ions colliding with the particle give off their charge to it and fly off as neutral molecules. Gentry and Brock,[207] apart from neglecting the image forces, made an incorrect assumption of a Boltzmann distribution of ion concentrations around the particle.

Since this distribution is due to molecular collisions and these are absent in the vicinity of the particle at $Kn_i \to \infty$, the formula derived in reference 207

$$\varphi_k = \pi a^2 \bar{V}_i n_\infty \exp(-i\varepsilon^2/akT) \tag{3.59}$$

is incorrect. The same error was committed by Liu, Whitby, and Yu.[208]

Keefe, Nolan, *et al.*,[209-211] who worked out the problem of a free trajectory of an ion moving in the electric field created by the particle, arrived at a correct solution. Neglecting the image forces, they obtained a formula for the ion flux to an oppositely charged particle, making an allowance for the Maxwellian ion velocities distribution the expression

$$\varphi_k = \pi a^2 \bar{V}_i n_\infty \left(1 - \frac{i\varepsilon^2}{akT}\right) \tag{3.60}$$

coinciding with eqn. (3.59) only at $i\varepsilon^2/akT \ll 1$. With an allowance for the image forces, the ion flux to an uncharged particle is

$$\varphi_k = \pi a^2 \bar{V}_i n_\infty \left[1 + \sqrt{\left(\frac{\pi\varepsilon^2}{2akT}\right)}\right]. \tag{3.61}$$

For the case of charged particles this problem can be solved only by a computational method.[211] The values of the "electrostatic" factor [contained in brackets in eqns. (3.60) and (3.61)] for this case are shown in Fig. 5 as the function of the particle radius. The curves in this figure have been calculated from root-mean-square velocities of ions. As shown by Keefe *et al.*, the curves allowing for ion velocity distribution can be obtained to a good approximation by shifting the curves shown in the figure to the right in such a way that the abscissa should increase 1.5 times. In their calculations Keefe *et al.* assumed the ion clusters to consist of eleven water molecules, i.e. they took $m_i/m_g \approx 7$.

The problem of particle charging at intermediate Kn presents enormous difficulties, and is far from being solved. On the basis of eqn. (3.59), Gentry and Brock,[207] using the Knudsen iteration method, obtained an expression for φ/φ_k of the form familiar to us. Liu, Whitby, and Yu[208] went even further: they assumed not only the Boltzmann distribution of ion concentrations near the particle, but also the Maxwellian (isotropic) velocity distribution and arrived at the conclusion that eqn. (3.59) is applicable at any Knudsen numbers.

All the other authors solved this problem by the boundary sphere method. The diffusional flux to the sphere was determined by means of eqn. (3.57) substituting for a the boundary sphere radius $a+\beta l_i$. The calculation of the free molecular flow to the particle from the boundary sphere was performed in different ways. Bricard[212] and Siksna[213] ignored the curvature of the ions' trajectory in the electric field of the particle. Keefe *et al.*[211] in their calculations used the results presented above for $Kn_i \to \infty$, i.e. they assumed that the ions come to the particle as if from infinity. In Fuchs's calculations,[214] the starting point of ion trajectories was assumed to lie on the boundary sphere. Natanson[215] chose for the boundary sphere such a sphere ion in which the ions that have undergone a collision inside it cannot leave. All the above-mentioned authors based their calculations on widely different values of l_i and $\bar{V}_i$, so that it is very difficult to compare their results. Since the boundary sphere method cannot give accurate results at all, it is hardly worth while to discuss the question which of the methods of calculation mentioned is more accurate.

In the foregoing we considered the process of charging and discharge of particles to be a quasi-steady one. However, in each act of particle recharging the ion concentration distribution around it must change. Evidently, for the charging process to be considered a quasi-steady one it is necessary that the relaxation time of the ionic atmosphere around the particle should be much less than the mean time interval between the recharging acts. As it follows from an analysis of the problem,[214] for this to be true the condition $a^3 n_\infty \ll 1$ should be observed at $a \geqslant 10^{-5}$ cm. At $a = 10^{-6}$ cm a more rigid condition $a^3 n_\infty \ll 0.01$ is to be satisfied. In practice these conditions can be considered as being observed both in laboratory and in the atmosphere.

As a rough calculation shows, under the conditions indicated it is also possible to neglect the electric field set up by a non-uniform ion distribution near a charged particle, i.e. by the shielding effect.

In the case when the particle charging occurs in a bipolarly ionized medium, ions of both signs can be considered to diffuse to the particle independently of each other. No matter what the initial particle charge distribution may be, a steady state distribution should be eventually established in aerosol, in which the number of particles with i charges of a given sign that have received in unit time one more charge of a like sign is equal to the number of particles with $i+1$ charges that have received one charge of the opposite sign. Let us assume for simplicity that the aerosol is monodisperse and that concentrations, mobilities, and other properties of ions of both signs coincide. Having determined by means of the formulas given above the fluxes of like and unlike ions to particles with i charges φ_i and φ_i', a series of equalities is written

$$2\varphi_0 N_0 = \varphi_1' N_1;\ \varphi_1 N_1 = \varphi_2' N_2; \ldots;\ \varphi_i N_i = \varphi_{i+1}' N_{i+1} \tag{3.62}$$

where N_i is the steady-state concentration of particles with i charges, and hence find the particle charge distribution (the coefficient 2 in the first equality is necessary because the uncharged particle can capture ions of both signs). It should be noted that this distribution does not depend on the ion concentration, but the rate of its establishment is proportional to n_∞.

For large particles ($a \geqslant 10^{-4}$ cm), when eqn. (3.57) is applicable and mirror forces can be neglected,

$$\varphi_0 = 4\pi D a n_\infty;\ \varphi_i = \frac{4\pi D a n_\infty \lambda_i}{\exp \lambda_i - 1};\ \varphi_i' = \frac{4\pi D a n_\infty \lambda_i}{1 - \exp(-\lambda_i)} \tag{3.63}$$

where $\lambda = |i\varepsilon^2/akT|$. It follows from eqn. (3.63) that

$$\varphi_i/\varphi_i' = \exp(-\lambda_i).$$

Multiplying the equalities (3.62),

$$N_i = N_0 \frac{\exp \lambda_i - 1}{\lambda_i} \exp\left(-\sum_1^i \lambda_k\right)$$

$$= N_0 \exp(-\varepsilon^2 i^2/2akT)[\exp(\lambda_i/2) - \exp(-\lambda_i/2)]/\lambda_i. \quad (3.64)$$

As shown by calculations, the factor $[\exp(\lambda_i/2) - \exp(-\lambda_i/2)]/\lambda_i$ for particles with $a \geqslant 10^{-4}$ cm and not too high charges (these being the overwhelming majority) is very close to 1. Discarding it,

$$N_i = N_0 \exp(-i\varepsilon^2/2akT), \quad (3.65)$$

i.e. the particle charge distribution is expressed by the Boltzmann equation. It should be noted that eqn. (3.65) is very accurate at $a \geqslant 10^{-4}$ cm, but with decreasing particle size it becomes less and less accurate. In connection with the method of particle-size determination from the percentage of charged particles under steady-state conditions, described in section 1.3, it should be pointed out that for $a = 10^{-5}$ cm this percentage, calculated according to Boltzmann, is less than the more accurately calculated one by approximately 2% (relative), for 3×10^{-6} cm by approximately 10%, for 10^{-6} cm approximately twice.

Keefe *et al.*[211] assumed *a priori* that the particle charge distribution conforms to the Boltzmann law, suggesting various explanations for deviations from it at small a. However, the Boltzmann formula is applicable only to equilibrium states. If ions attached to particles tore away from them as ions, i.e. if there existed a "detailed" equilibrium, the Boltzmann formula would be undoubtedly applicable, but in reality we have to deal with a steady rather than with an equilibrium state.

As was already stated in the introduction, the number of elementary charges acquired by HDA particles in a bipolar ionic atmosphere is very small. For illustration Table 2 shows the HDA particle charge distribution in the steady state calculated by Fuchs[214] (to within hundredths).

TABLE 2. STEADY-STATE HDA PARTICLE CHARGE DISTRIBUTION IN A BIPOLARLY IONIZED ATMOSPHERE

a (cm)	10^{-7}	3×10^{-7}	10^{-6}	3×10^{-6}	10^{-5}
Fraction of uncharged particles	0.99	0.95	0.76	0.51	0.29
Fraction of particles with 1 charge	0.01	0.05	0.24	0.45	0.44
Fraction of particles with 2 charges	—	—	—	0.04	0.20
Fraction of particles with 3 charges	—	—	—	—	0.06
Fraction of particles with 4 charges	—	—	—	—	0.01

In the presence of external electric field, the diffusional charging of particles is superimposed by the charging due to flux of ions moving in this field. In the case of unipolar ionization, this flux is expressed by Pauthenier's formula,

$$\varphi_E = 3\pi EnBa^2(1 - i\varepsilon/3Ea^2)^2, \quad (3.66)$$

where E is the field strength. To compare this flux with the diffusional flux let us consider the charging of a neutral particle of $a = 10^{-5}$ cm. In this case, it is possible without appreciable error to express the diffusional flux by eqn. (3.63) for φ_0. From this formula, as well as from eqn. (3.66), it is easy to find that

$$\varphi_E/\varphi_0 = 0.75\varepsilon Ea/kT = 0.9\cdot10^4 Ea \quad (3.67)$$

or, for $a = 10^{-5}$ cm, expressing E in v/cm,

$$\varphi_E/\varphi_0 = 3\times 10^{-4}E. \tag{3.68}$$

For smaller particles and for charged ones, this ratio is still less. Thus the rate of the "drift" charging of HDAs becomes equal to that of the diffusional charging only in rather strong fields (over 3000 v/cm), e.g. in electric filters, whereas in atmosphere it can be neglected.

Passing to the experimental data on the kinetics of charging of aerosol particles by gas ions, let us note that most of the measurements were performed on relatively large particles at atmospheric pressure, i.e. at small Kn. Flanagan[(216)] measured the decrease of the steady-state concentration of gas ions in air as affected by polydisperse aerosol produced by heating nichrome, with $\bar{a}$ from 0.01 to 0.13 μm, i.e. at $Kn \geqslant 1$. Assuming that in this case the particles received not more than one elementary charge (which is true only for the lower part of this range), Flanagan determined from these measurements the sum of the quantities φ_0 and φ_1', viz. $\varphi_0+\varphi_1' = 0.58a$. Such a relation at $Kn \geqslant 1$ is in poor agreement with theory.

Liu, Whitby, and Yu[(217)] measured the charging kinetics of monodisperse aerosols of dioctylphthalate with $a = 0.32$ and 0.67 μm by unipolar ions at pressures of 0.03–1 atm, i.e. at $Kn_i \leqslant 1$, and found that the curve of particle charge versus time is in good agreement with eqn. (3.59) in which Kn does not appear, but which the above-mentioned authors considered to be valid at any Kn. For this purpose, they had to assume, however, the mass of ions to be 16 times as large as that of air molecules. As has been already pointed out, eqn. (3.59) is correct at all values of Kn and it is incomprehensible how such agreement could have been attained. It should be noted, however, that the charging experiments carried out at low pressure cannot simulate accurately the process of charging of HDA particles owing to the effect of image forces, the magnitude of which depends on the particle size rather than on Kn_i.

A number of authors measured the ratio between the charged and uncharged particles in the steady state in a bipolar ionic atmosphere and thence calculated the ratio

$$\varphi_i'/\varphi_0 = 2N_0/N_i. \tag{3.69}$$

These experiments were carried out on polydisperse aerosols and can hardly be used for verification of theoretical computations. It is interesting to note that Pollak and Metnieks,[(218)] who used aerosols with $\bar{a} = 1-4\times 10^{-6}$ cm prepared by heating nichrome, found that, in agreement with theory, the experimental values of the ratio N_i/N_0 are higher than those calculated by means of the Boltzmann formula at $\bar{a} < 1.35\times 10^{-6}$ cm and lower for larger particles. These authors showed that this discrepancy was due to polydispersity of aerosols which the larger the particles distort the results the more.

3.6. Thermophoresis and diffusiophoresis

In the treatment of these complicated phenomena the range $Kn \gtrsim 1$ will be the limit.

A particle in an inhomogeneous gas medium is acted upon by a force which is called thermophoretic if it is due to the presence of a temperature gradient in the medium, and diffusiophoretic if there is a mixture composition gradient. The force acting upon the particle surface S is determined by integral equations

$$F_{Tk} = -\int dS \sum_{\pm} \int m_g[\overline{V}_g\times \overline{V}_g]f^{\pm}\, dv_g, \tag{3.70}$$

where f^+ and f^- are the distribution functions of gas molecules whose velocity projection to the direction of the temperature or concentration gradient is positive or negative respectively, and $\bar{V}_g$ is the mean velocity of gas molecules. The gradient appears in this equation in its implicit form in terms of the functions f^+ and f^- depending on the co-ordinates.

The reasons for the motion of aerosol particles due to the temperature gradient differ for $Kn \gg 1$ and $Kn \ll 1$. At $Kn \gg 1$ this phenomenon is exactly similar to thermal diffusion. In the case of $Kn \ll 1$ the motion of aerosol particles is due to the existence of a temperature gradient inside the particle itself and the tangential motion of gas along the non-uniformly heated particle surface.

Unlike other transfer phenomena, it is impossible in calculating the thermophoretic force even at $Kn \gg 1$ to use the zero approximation for f, since the gas is in a non-equilibrium state.

The theory of thermal diffusion in gases was used in solving the problem of thermophoresis in a free molecular regime by Einstein,(219) Cawood,(220) and Clusius,(221) and in a more rigorous form by Derjaguin and Bakanov(223) as well as by Waldmann.(222) In the last two papers use was made of the Chapman–Enskog theory in Lorentz approximation, the mass of aerosol particles being considered large compared to the mass of gas molecules, and the role of the collisions of molecules with one another negligible compared to that of their collisions with the particle surface. If we assume a fraction of molecules $(1-\varphi)$ to be reflected specularly and the rest diffusely, i.e. with a Maxwellian distribution corresponding to the temperature of the particle surface, the thermophoretic force does not depend on φ and is expressed by

$$F_{Tk} = -(32/15)(a^2\chi_g \nabla T/v_g), \tag{3.71}$$

where ∇T is the temperature gradient and χ_g is the heat conductivity of gas. In the case of polyatomic gas, the value of the translational component of heat conductivity $\chi_{gt} = 15k\eta/4m_g$, where η is viscosity, should be substituted into eqn. (3.71). Mason and Chapman(224) assumed all molecules to be reflected elastically, partly specularly and partly φ retaining their absolute velocity, but in random directions. They obtained a $1+4\pi/9$ times larger value of F_{Tk} than that in eqn. (3.71). Monchik, Yu, and Mason(225) using the model of inelastic collisions developed by Wang Chang and Uhlenbeck(226) substantiated the applicability of eqn. (3.71) in the case of polyatomic gases.

The theory of thermophoresis for intermediate Kn was developed by Brock,(227) who used Bhatnagar's method in conjunction with Knudsen iteration. According to Brock, for a one-component polyatomic gas the thermophoretic force is

$$F_T = F_{Tk} \exp\{-[0.06+0.09\alpha_m+0.28\alpha_m\alpha_T\chi_a/2\chi_g]Kn^{-1}\}, \tag{3.72}$$

where F_{Tk} is given by eqn. (3.71), χ_g and χ_a are the heat conductivities of gas and particle respectively, and α_m is the coefficient of momentum accommodation. This equation is equivalent to the empirical formula obtained by Schmitt,(228)

$$F_T = F_{Tk} \exp(-b/Kn) \tag{3.73}$$

if it is assumed that for monatomic gas

$$b = 0.06+0.37\alpha_m-0.09\alpha_m\alpha_T\chi_a/2\chi_g, \tag{3.74}$$

and for polyatomic gas

$$b = 0.06+0.37\alpha_m-0.28\alpha_m\alpha_T\chi_a/2\chi_g^*, \tag{3.75}$$

where χ_g^* is the total heat conductivity of polyatomic gas and α_T is the coefficient of thermal accommodation.

The values of b according to reference 229 are given in Table 3.

TABLE 3. THE VALUES OF THE CONSTANT b IN EQN. (3.73) FROM DIFFERENT EXPERIMENTAL DATA

Reference	Aerosol substance	Gas medium	b	α_m
228	0.2	Ar	0.39	0.80
228	0.2	N_2	0.38	0.81
228	0.2	CO_2	0.36	0.80
228	0.2	H_2	0.22	0.79
240	Tricresylphosphate	Air	0.41	0.88
241	NaCl (amorphous)	Ar	0.47	0.91
234	NaCl (crystalline)	Air	0.46	0.91

Recently, Dwyer[230] developed the theory of thermophoresis in a thirteen-moment approximation, which he believes to be applicable over the whole range of Kn. This point of view is unjustified since the thirteen-moment approximation was developed for the slip regime, i.e. for $Kn < 1$. In fact the application of this approximation to the description of the coagulation kinetics[231] gave results essentially different from accurate theory already at $Kn \approx 1$. In the range $Kn \to \infty$, Dwyer's result does not turn into the accurate formula (3.71), but predicts a monotonic decrease of F_{Tk} with increasing Kn. Dwyer's conclusion that F_T should have a negative value at the values of the coefficient of mechanical accommodation $0.9 < d_m < 1.0$ is at variance with all known experimental data.[232]

Gardner[233] studied the problem of thermophoresis of an evaporating particle contained in its own rarefied vapor in the absence of gas-carrier and showed that in this case the thermophoretic force should be practically the same as in the absence of evaporation.

An experimental investigation of the thermophoresis of silicone oil particles in Ar, N_2, CO_2, and H_2, partly in the range of $Kn \approx 1$ was carried out by Schmitt.[228] The experiments consisted in observing the migration of a single drop in a Millikan's condenser with plates thermostated at various temperatures. The particle radii were 0.7–1.2 μm and pressure 10–760 mm, so that the range of $Kn \approx 0.05$–3.3 was covered. At $Kn \approx 3.3$ F_T is strictly proportional to a^2, which corroborates the theory qualitatively. In the range $Kn \approx 0.2$ the experimental data are approximated by eqn. (3.73); for all gases except H_2, $b = 0.38$. At $Kn \approx 3$ the ratio of the values of F_T calculated from eqn. (3.71) to experimental values in Ar, H_2, CO_2, and H_2 is equal to 1.05, 1.21, 1.22, and 1.38 respectively, the experimental accuracy being the least in the last-mentioned case. A similar method was used by Schadt and Cadle[234] for sodium chloride and mercury particles with particle radii 0.22–1.15 and 0.096–0.145 μm respectively within the range of $0.2 < Kn < 4.0$. The measurements showed the independence of F_T of the heat conductivity of particles, i.e. they confirmed the theoretical conclusion. However, the measured values were closer to Cawood and Einstein's theory than to calculations according to Derjaguin and Waldmann.

Two particular cases should be distinguished in the treatment of diffusiophoresis: (1) equimolecular counterdiffusion of two components, and (2) the diffusion of one component through the second stationary component. The theory of diffusiophoresis at large Kn was developed by Waldmann[235] and Derjaguin and Yalamov.[236] In case (1) the force acting upon the particle is

$$F_{dk} = a^2 n_g (2\pi kT)^{\frac{1}{2}} D_{12} (\nabla x_1)_\infty [m_{g_2}^{\frac{1}{2}}(1+\varphi_2\pi/8) - m_{g_1}^{\frac{1}{2}}(1+\varphi_1\pi/8)], \quad (3.76)$$

where D_{12} is the coefficient of interdiffusion, X_1 is the mole fraction of component 1, and $(\nabla X_1)_\infty$ is the value of the concentration gradient of component 1 at infinite distance from the particle.

For case (2)

$$F_{dk} = -\tfrac{8}{3}a^2 n_g(2\pi m_{g_1} kT)^{\frac{1}{2}}\left(1+\frac{\varphi_1\pi}{8}\right)\frac{D_{12}}{x_2}(\nabla x_1)_\infty. \tag{3.77}$$

Diffusiophoresis in a transient regime in a two-component mixture was studied by Brock[237] by means of Bhatnagar's model, so that the applicability of the theory was limited to the case of molecules with similar size and mass values. For case (1) and $x_1 \gg x_2$, if it is assumed that $\varphi_1 = \varphi_2 = 1$:

$$F_d = F_{dk} + \left\{\left[\frac{d_1 m_g^{\frac{1}{2}}}{d_{12}^2} - \left(\frac{2m_{g_1} m_{g_2}}{m_{g_1}+m_{g_2}}\right)^{\frac{1}{2}}\right]\frac{0.311a}{l_g}\right\}, \tag{3.78}$$

where d_1 and d_2 are effective diameters of molecules, $d_{12} = (d_1+d_2)/2$.

For case (2) and $x_1 < x_2$,

$$F_d = F_{dk}\{1-0.071[2m_2/(m_1+m_2)]^{\frac{1}{2}} Kn^{-1}\}. \tag{3.79}$$

In eqns. (3.78) and (3.79) the values of F_{dk} are taken from eqns. (3.76) and (3.77) respectively.

At large and intermediate Kn, diffusiophoresis was experimentally investigated by Waldmann and Schmitt.[238] The measurements were performed on silicone oil drops at 20° for different binary gas mixtures. For case (1) the results agreed with eqn. (3.76) in the mixtures CO_2/C_3H_8, N_2/C_3H_8, N_2/C_2H_4, N_2/C_2H_6, N_2/O_2, and N_2/Ar. Brock[239] compared the calculated values of the ratio $v_d/D_{12}(\nabla x)_\infty$ (v_d is the velocity of diffusiophoresis) at $Kn \to \infty$ with Waldmann and Schmitt's results and obtained good agreement for mixtures of nitrogen with oxygen and hydrocarbons listed above. For the mixtures N_2/Ar and CO_2/C_3H_8 no agreement between theory and experiment was obtained owing to a strong difference in the diameters of the components, which precludes the use of a simplified relaxation model of the kinetic equation for a binary mixture. For case (2) Brock obtained a fair agreement for asymptotic values of the ratio $V_d x_2/D_{12}(\nabla x_1)_\infty$, found from theory and experiment at $Kn \to \infty$ in H_2O/N_2 mixtures.

Brock[239] pointed out that there should exist one more form of the motion of aerosol particles—"photodiffusiophoresis", resulting from the change in the physical or chemical equilibrium between particle and gas medium under the action of electromagnetic radiation. For the usual photophoresis, which is known to be due to a non-uniform particle heating under the action of radiation, Brock suggested the name "photothermophoresis".

3.7. Coagulation of HDAs

As has been shown by Smoluchowski, the problem of Brownian coagulation is mathematically equivalent to the problem of vapor condensation on a particle. Therefore all that has been said in section 3.2 regarding mass transfer is applicable to coagulation, the only difference being that in the case of coagulation the Knudsen number should be determined in a somewhat different manner: first of all, here the apparent (or effective), mean free path of aerosol particles l_p, i.e. the average distance at which the correlation with the initial direction of the particle motion disappears, should be used instead of the mean free path of gas molecules. The value of l_p is given by the relation $l_p = \bar{v}_p\tau$, where

$\bar{v}_p = (8kT/\pi m_p)^{\frac{1}{2}}$ is the mean thermal velocity of the particle, $\tau = m_p B$ is its relaxation time, B its mobility, and m_p its mass. In the case of coagulation, l_p should be substituted by a corresponding quantity for the relative motion of two particles, i.e. by $l_p\sqrt{2}$. Finally, instead of the particle radius, the "absorbing sphere" radius $r = 2\alpha$ should be used. By virtue of the above, the aerosol Knudsen number is expressed as $Kn_p = l_p(\sqrt{2}/2)$. The values of l_p and Kn_p for spherical particles of unit density suspended in air under normal conditions are given in Table 4.

TABLE 4. COAGULATION OF HDAS (WITHOUT ALLOWING FOR MOLECULAR FORCES)

Particle radius (Å)	10	20	50	100	200	500	1000
l_p (Å)	659	468	30	22	164	124	113
Kn_p	46.4	16.5	4.24	1.56	0.58	0.176	0.08
Coagulation constant: $K \times 10^{10}$ cm³/sec							
Smoluchowski's equation	323	162	65.8	34.0	18.0	8.57	5.56
Free molecular	4.42	6.26	9.90	14.0	19.8	31.3	—
Sahni's theory	4.39	6.14	9.67	12.0	11.0	7.1	5.14

It is evident that at large Kn_p the coagulation constant should be expressed by the formula for the number of collisions between gas molecules, i.e. the aerosol particles can be treated as large gas molecules, in which case the probability of collisions between molecules is not affected by the prssence of another light component (Lorentz gas),

$$K_k = 2\sqrt{(2)}\pi a^2 \bar{v}_p = 4(3akT/\rho)^{\frac{1}{2}}, \tag{3.80}$$

where ρ is the particle density. The applicability of this expression was rigorously substantiated by Hidy and Brock,[242] who solved the kinetic equation for the Lorentz gas. The expression for the coagulation constant obtained by them takes account of the electrostatic interaction between particles, the effect of external fields, and the presence of the velocity gradient, but, unfortunately, does not contain the terms with molecular forces, which are undoubtedly of importance in HDA coagulation since their particle size is smaller or comparable with the phonon wavelength. The potential of intermolecular interaction for two spherical particles of the same radius is described by Hamaker's equation.

$$U(R) = \frac{A}{6}\left[\frac{2a^2}{R^2} + \frac{2a^2}{(R^2-4a^2)} + \ln\left(1 - \frac{4a^2}{R^2}\right)\right], \tag{3.81}$$

where R is the distance between the particle centers $A = \pi g^2\beta/6$, g is the number of molecules in unit volume of substance and β is the London constant. From the equation of momenta it is easy to obtain the expression

$$b^2 = R_m^2[1-U(R_m)/E], \tag{3.82}$$

where b is the impact parameter (Fig. 6), R_m is the minimum distance between the particle centres, and E is the total energy of the particle. The calculation by the method of successive approximations shows that in the range $2a < R_m < \infty$ the function $b(R_m)$ should have only one maximum b_m. At $b < b_m$, R_m becomes imaginary, i.e. there exists no minimum distance from the trajectory to point 0, and the particles must collide. At $b > b_m$, R_m has a real value, and this corresponds to the case shown in Fig. 6—the absence of collision. For two moving particles, E is the kinetic energy of their relative motion at $U = 0$. Therefore the mean value of E is $3kT$. In the absence of molecular forces, $b_m = 2a$, i.e. the molecular forces increase $b_m^2/4a^2$ times the particle collision cross-section and, hence, the coagulation constant.

By an approximate solution of the kinetic equation, Hidy and Brock(243) determined the coagulation constant at intermediate Kn_p. In their deduction each particle was considered as being a source of molecules with a disturbed distribution function. Colliding with the surface of another particle, these molecules set up a repulsive force acting along the centre axis. The formula obtained by Hidy and Brock is very complex. At $Kn_p \geqslant 5$ the correction

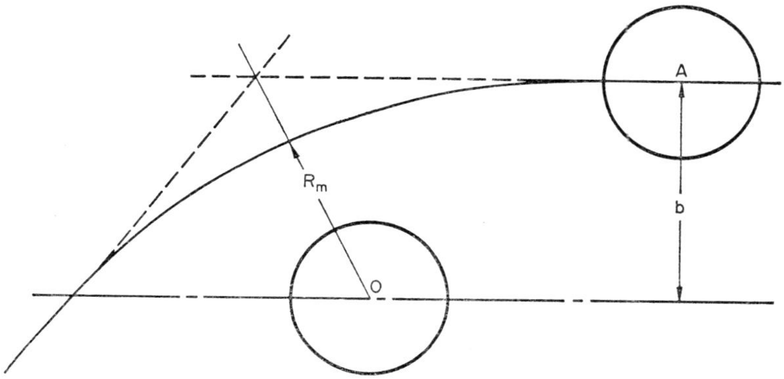

FIG. 6. Particle trajectory in the central attraction field.

factor for the free molecular formula (3.80) is very near to unity. With decrease of Kn_p this factor diminishes somewhat and becomes by fractions of 1% less than unity, but at $Kn_p \approx 2$ it begins to grow fast with decreasing Kn_p. Thus at $Kn_p \simeq 1$, the Hidy and Brock formula predicts larger values of the coagulation constant than the free molecular one—a result which is inconsistent with its physical sense.

Apparently the most accurate data on the value of the coagulation constant can be obtained by means of the method described in section 3.2, using eqn. (3.24). The values of the coagulation constant calculated by this method are given in Table 4. For comparison there are also given the values calculated by means of eqn. (3.80) and the Smoluchowski equation, which some of the authors dealing with HDAs still believe to be applicable.

For the determination of the coagulation constant Astakhov(231) used the thirteen-moment approximation. But the formula obtained by him gives correct values only for the range $Kn_p \approx 1$.

The problem of the effect of molecular forces on coagulation at intermediate Kn_p is very complicated since it is necessary to take into consideration the effect of the retardation of molecular forces at distances comparable to the phonon wavelength. No theoretical computation taking into consideration the molecular forces at $Kn_p \approx 1$ has been suggested so far.

Recently, Friedlander and Wang(244) found that in the case of aerosol coagulation under the conditions when the coagulation constant depends on Knudsen number, no self-preserving particle size distribution is established.

There are very few experimental studies available on HDA coagulation. The main difficulty in these studies is a considerable loss of particles on the walls of the apparatus and communications due to their large diffusion coefficient. For this reason, the classical approach to the study of coagulation—the determination at regular intervals of the concentration of particles sampled from a chamber—is inapplicable in the case of HDAs. Therefore it is no wonder that O'Connor[245] and Quon,[246] who used this method, obtained theoretically inexplicable results, viz. that with decreasing particle size in the range of intermediate Kn_p the coagulation constant increases monotonically and even faster than it should follow from the Smoluchowski equation. O'Connor worked with room condensation nuclei with undefined particle composition and shape. Quon used aerosols which were products of incomplete combustion of hydrocarbons. Though on account of a low particle concentration, these aerosols could be considered to consist of individual particles rather than on aggregates, the question as to their form and degree of polydispersity remained unclarified. In reference 101 only qualitative information on the nature of HDA coagulation was obtained.

Fuchs and Sutugin[98] used a flow method of investigation of HDA coagulation. A laminary stream of aerosol with a large previously measured particle concentration was caused to flow through a wide tube, and the ratio of HDA concentrations at entrance and exit was measured with a nephelometer. Before nephelometric measurements, HDA particles were enlarged up to approximately 0.5 μm by condensing on them dibutylphthalate vapors in a PSM (see p. 19). In some experiments the diffusion losses of particles in the tube were determined, for which purpose diluted aerosol was passed through it. The initial particle concentration was determined by means of an ultramicroscope after dilution of HDAs in a specially designed diluting unit and enlarged in the PSM. The coagulation constants for monodisperse sodium chloride aerosols of mean particle radius 25 and 45 Å, determined by this method, were equal to 13.4×10^{-10} and 15.1×10^{-10} cm^3/sec respectively. The r.m.s. error in measurements was 11 %. The correction factor $b_m^2/4a^2$ to the free molecular expression for the coagulation rate, due to molecular forces, was calculated. The use of the London constant computed by Meyer[247] from the sodium chloride ultraviolet absorption bands gave the value of the correction factor 2.19. With account taken of this correction factor, the calculated values of the coagulation constant were 10.2×10^{-10} cm^3/sec for $\bar{a} = 25$ Å and 13.4×10^{-10} cm^3/sec for $\bar{a} = 45$ Å. At any rate, the last-mentioned value agrees well with experimental data. The discrepancy between experiment and theory can be accounted for by a certain aerosol polydispersity and inaccurately determined value of the London constant.

The method described was also used in reference 248 for measuring the coagulation constant of monodisperse dioctylcebacate aerosols with $\bar{a} = 100$ Å. In these experiments, Kn_p was 1.54. The experimental value of the coagulation constant was 13.4×10^{-10} cm^3/sec, i.e. somewhat greater than the calculated value given in Table 4. In this case, the discrepancy can be completely accounted for by the experimental error and the effect of molecular forces.

Stockham[13] calculated the coagulation growth of silver particles and found that his experimental results could be approximated by calculated data if the coagulation constant was taken to be 2–10 times larger than the free molecular value. The discrepancy between theory and experiment increased with the concentration of vapor from which aerosol was formed. This is evidently due to the formation of branched-chain aggregates in the case of coagulation of aerosols with large weight concentration. At low concentrations, as it follows from Stockham's electronic photomicrographs, the aggregates were compact and the coagulation constant was approximately twice as high as the free molecular value, which agrees well with the effect of molecular forces on coagulation of silver particles (see p. 8).

From the above results it can be concluded that there is a satisfactory agreement between theory of HDA coagulation and experiment.

The theory of coagulation of charged aerosols in molecular and transient regimes is mathematically equivalent to the theory of charging of aerosol particles. In the free molecular regime the value of the coagulation constant of charged aerosols can be calculated by means of an expression obtained by Hidy and Brock.(242) For intermediate Kn_p such calculations were performed only using the boundary-sphere method.(249)

3.8. Capillary effects and structure of HDA particles

In addition to being peculiar in their behavior with respect to transfer phenomena and in their optical and electrical properties, HDAs also possess some specific thermodynamic properties, the most important of which—the vapor pressure increase—is taken account of in the well-known Kelvin equation

$$\ln(p/p_0) = 2M\sigma/\rho R T_a, \tag{3.83}$$

where p_0 is the equilibrium vapor pressure over a plane surface, p is the vapor pressure over a droplet of radius a, σ is the surface tension, and ρ is the density of the substance.

The Kelvin effect becomes apparent beginning with the droplet size 0.1–0.3 μm, and increases rapidly with decreasing Kn, which is one of the reasons of HDA instability or even of the impossibility to prepare them from substances with not too low vapor pressure at room temperature. An indirect experimental verification of eqn. (3.83) was carried out by La Mer and co-workers(95, 250) for sulfuric acid and dibutylphthalate droplets. Recently, the evaporation rate of liquid lead and bismuth solid-gold particles was directly studied by means of an electron microscope on a carbon support.(251) These experiments carried out on particles of 100–500 Å also confirmed the validity of the Kelvin equation.

The presence of increased pressure over a surface of positive curvature is sometimes called the capillary effect of the first kind. The capillary effects of the second kind are the dependence of surface tension on the surface curvature and the dependences of the pressure in the droplet and its density on the size.

Tolman(252) was the first to show that the existence of the $\sigma(a)$ dependence is a logical sequence of Gibbs thermodynamics. He obtained an integral equation

$$\ln(\sigma/\sigma_0) = \int_0^\infty \frac{(2\delta_0/a^2)[1+(\delta_0/a)+(\delta_0/a\sqrt{3})^2]}{1+(2\delta_0/a)[1+(\delta_0/a)+(\delta_0/a\sqrt{3})]}\,da, \tag{3.84}$$

where σ_0 is a quantity of the order of the interatomic distance. To the first approximation,

$$\sigma/\sigma_0 = 1-2\delta_0/a. \tag{3.85}$$

Here σ_0 is the surface tension of a plane surface. The theory of surface tension of curved surfaces was further developed in references 253–255, in which the inaccuracy of Tolman's deduction due to the equating of the pressure in the drop and in the macroscopic liquid volume was eliminated. In those studies the value of the pressure was calculated by means of various equations of state. For δ an integral equation

$$\int_{-\infty}^{+\infty} (R+\delta_0)[p_0 - p_{xx}(R)]\,dR = 0 \tag{3.86}$$

was obtained, where R is the radial coordinate and p_{xx} is the pressure tensor component describing the forces acting in the direction perpendicular to R upon a spherical area element.

The $\sigma(a)$ dependence is of great importance for the theory of spontaneous condensation. However, as was pointed out by Kirkwood and Buff,[(256)] this dependence cannot be taken into consideration without allowing for the density decrease of fine particles. The radial density distribution and surface tension of particles of diameter 3–25 times larger than the molecule size was calculated by Plesner[(257)] by means of an equation of state for solid spheres with the attraction potential inversely proportional to the sixth power of the distance. The result obtained by Plesner proved to be surprising—the surface tension of such droplets does not depend on their size and is equal to $(1/3)\sigma_0$. This means that in a certain size range—a rather narrow one inasmuch as the experimental dependence $\sigma(a)$ has not been observed up to $a = 10^{-6}$ cm—the value of σ should decrease sharply. This seems improbable. According to Plesner, for very small droplets the density does not reach its macroscopic value even in their center.

Shcherbakov[(257,258)] pointed out that for very small particles the free energy excess should be referred to their volume rather than to the surface alone. He also found that substituting merely the $\sigma(a)$ dependence into the Gibbs–Kelvin equation does not give a correct result, it being necessary to introduce into this equation a term containing $d\sigma/da$. Earlier, a similar point of view was expressed by Gibbs and Kelvin, but it was purely intuitive. Shcherbakov considers the surface tension as the free energy excess referred to as the unit surface of particle. According to Shcherbakov, the $\sigma(a)$ dependence is of the form

$$\sigma = \sigma_0\left(1 - \frac{2}{ka} + \frac{2}{k^2a^2}\right), \tag{3.87}$$

where $(1/k) \approx 2$–5 Å. Owing to increasing influence of the third term with decreasing a, the function $\sigma(a)$ has a minimum, and at very small a increases with decreasing size. This result is at variance both with the results in references 252–255 and with that obtained by Bellemans[(260)] in the generalization of Mayer's theory of the equation of state to systems containing an interface of arbitrary curvature.

A monotonically increasing function $\sigma(a)$ is predicted also by the theory of surface tension based on the assumption that the free energy density is a function both of concentration and concentration gradient of molecules, developed by Cahn and Hilliard[(261,262)] as well as by Hart.[(263)] According to this theory, at high supersaturation in dense vapors, liquid phase particles have a very diffuse surface, with their density differing but insignificantly from vapor density. In developing this theory, Hart[(264)] came to a very important conclusion. In the Gibbs theory of surface tension the quantity σ is invariant with respect to the choice of interfacial boundary in the transition region between the phases. For a curved surface, however, this invariancy does not hold, and the calculated value of σ depends on the position of this boundary. In this case the position of the boundary and the value of σ can be found from the condition $d\sigma/dR = 0$, where R is the radial coordinate. Hart found that under certain conditions the function $\sigma(R)$ can have no minimum, i.e. notion of surface tension becomes senseless.

Okuyama and Zung[(267)] studied theoretically the dependence of the evaporation coefficient on curvature. They pointed out that the existence of this dependence is due to the necessity to take into consideration the work of surface increase of a droplet upon addition of another molecule to it. Taking account of this fact, they obtained

$$\alpha_c = \alpha_{c0} \exp(-\varepsilon_0/kT), \tag{3.88}$$

where α_{c0} is the condensation coefficient on a plane surface and ε_0 is the activation energy of adhesion to the surface at 0°K. According to the above authors, ε_0 is equal to the surface energy of a droplet divided by the number of molecules in it g. Then,

$$\alpha_c = \alpha_{c0} \exp(-4\pi a^2 \sigma / gkT). \tag{3.89}$$

A decrease of α_{c0} calculated using σ_0 becomes appreciable starting from $a \approx 100$ Å and for water at 0°C and $a \sim 1$ Å decreases 10^8 times compared to α_{c0}. Evidently, taking into consideration the dependence $\sigma(a)$ should significantly decrease this effect if not eliminate it altogether.

The decrease of the melting point of fine particles is described by G. G. Thomson's formula

$$\Delta T_m / T_m = 2\sigma V_e / La, \tag{3.90}$$

where ΔT_m is the decrease of the melting point, T_m is the macroscopic value, V_e is the molar volume of the liquid, and L is the melting heat. Lower melting temperature is responsible for the ability of HDA particles to preserve an amorphous state for a long period of time and their ability to coalesce or sinter upon contact. The results of a whole number of studies(32, 36, 48) show that the particles of condensation HDAs are usually amorphous. This fact is consistent with the Ostwald rule and seems natural considering that the free energy of formation of a liquid phase nucleus from vapor is lower than that of crystalline nucleus. Crystallization of aerosol deposits of aluminum(32) and of some metal oxides(36) requires heating or X-ray radiation. It was shown in reference 45 that in dry air sodium chloride particles of radius equivalent to 10^{-7} cm remain in the amorphous state for a long time. In moist air, however, large particles recrystallize within several hours and fine ones are crystalline from the outset.

The ability of HDA particles to coalesce or sinter upon contact has a very strong effect upon their coagulation nature. It is known that for many substances Tamman temperature at which solid bodies begin to sinter is equal to 0.6 of their absolute melting temperature.(265) The existence of the dependence of Tamman temperature on the particle size was theoretically proved by Higuchi.(266)

The authors of reference 38 report a sharp decrease in the specific area of a nickel aerosol deposit with particle size 30 Å at room temperature observed after several hours. It is interesting to note that the authors failed to detect the effect on this process of the composition of the atmosphere in which the experiments were carried out. Coagulation of silver aerosol with particle size approximately 100 Å involves sintering and coalescence of particles already at 120–150°,(6) i.e. at a much lower than normal Tamman temperature. A decrease of this temperature by several hundred degrees for HDA deposits of some high-melting substances was observed in reference 37. It is the ability of particles to sinter during coagulation that limits the possibility of preparing highly dispersed powders by way of aerosol state.(6)

REFERENCES

1. Hirth, J. and Pound, G., *Evaporation and Condensation*, Pergamon Press (1964).
2. La Mer, V., *Ind. Engng. Chem.* **44**, 1229 (1952).
3. Courtney, W., *Heterogeneous Combustion*, New York, Academic Press, vol. 15, p. 677 (1964).
4. Christiansen, J., *Acta chem. scand.* **5**, 676 (1951).
5. Reed, S., *J. Chem. Phys.* **20**, 208 (1952).
6. Sutugin, A. G. and Fuchs, N. A., *J. Colloid Interface Sci.* **27**, 216 (1968).
7. Hidy, G. and Lilly, D., *J. Colloid Sci.* **20**, 867 (1965).
8. Hidy, G., *J. Colloid Sci.* **20**, 123 (1965).
9. Friedlander, S. K. and Swift, D., *J. Colloid Sci.* **19**, 621 (1964).
10. Friedlander, S. K. and Wang, C., *J. Colloid Interface Sci.* **22**, 126 (1966).
11. Beeckmans, J., *Can. J. Chem.* **43**, 2312 (1965).
12. Rosinski, J. and Snow, J., *J. Meteorol.* **18**, 736 (1961).
13. Stockham, J., *Microscope*, **15**, 102 (1966).
14. Stogryn, D. and Hirschfelder, J., *J. Chem. Phys.* **31,** 1531 (1959).
15. Griffin, J. and Sherman, P., *AIAA Jl.* **3**, 1813 (1965).
16. Hill, P., Wittig, H. and Demetry, E., *J. Heat Mass Transfer*, **85**, 303 (1963).
17. Wegener, P., *Phys. Fluids*, **7**, 352 (1964).
18. Tunitskii, N. N., *Zh. fiz. Khim.* **15**, 1061 (1941).
19. Corner, J., *Proc. Roy. Soc. A.* **211**, 417 (1952).
20. Oswatitsch, K., *Z. angew. Math. Mech.* **22**, 1 (1942).
21. Buykov, M. V. and Bakhanov, V. P., *Kolloid. Zh.* **29**, N6 (1967).
22. Dunham, S., *J. Rech. Atmos.* **2**, 331 (1966).
23. Tessner, P. A., *Discuss. Faraday Soc.* **30**, 170 (1960).
24. O'Connor, T., *et al., Geofis. Pura Appl.* **42**, 102 (1959).
25. Lodge, J. and Tufts, B., *J. Colloid Sci.* **10,** 256 (1955).
26. Goyer, G. and Pidgeon, F., *J. Colloid Sci.* **11**, 697 (1956).
27. Megaw, W. and Wiffen, R., *Proceedings 1st National Conference Aerosols, Prague, 1962*, Lublice, p. 511 (1965).
28. O'Connor, T. and Roddy, A., *J. Rech. Atmos.* **2**, 239 (1966).
29. Goldsmith, P. and May, F., *Nature*, **210**, 475 (1966).
30. Gen, M., *et al., Dokl. Akad. Nauk SSSR*, **127**, 366 (1959).
31. Gen, M., *et al., Zh. Tekh. Fiz.* **23**, 1407 (1959).
32. Gen, M. and Petrov, Ju., *Priborȳ i Tekh. Eks.* **4**, 162 (1963).
33. Kimoto, K., Kamija, J. and Nanoyama, M., *Jap. J. Appl. Phys.* **2,** 702 (1963).
34. Turkevitch, J., *Fundamental Phenomena Material Science*, Plenum Press, New York, vol. 3, p.195 (1966).
35. Sutugin, A. G. and Fuchs, N. A., *Zh. Prikl. Khim.* **39**, N3 (1969).
36. Harvey, J. and Matthews, H., *Discuss. Faraday Soc.* **30**, 113 (1960).
37. Holmgren, J., Gibson, J. and Sheer, C., *J. Electrochem.* **111**, 362 (1964).
38. Selover, T., *AIChE Jl.* **10**, 79 (1964).
39. Sofronov, N., *Kolloid. Zh.* **29**, 327 (1967).
40. Fuchs, N. A. and Sutugin, A. G., *Kolloid. Zh.* **25**, 487 (1963).
41. Spurny, K. and Hampl, V., *Coll. Czech. Chem. Commun.* **30**, 507 (1965).
42. Sutugin, A. G., *Kolloid. Zh.* **27**, 789 (1965).
43. La Mer, V., *Air Pollution* (Proc. US Techn. Conf. Air Pollution), Los Angeles, p. 607 (1952).
44. Matiević, E. and Kerker, M., *Discuss. Faraday Soc.* **30**, 178 (1960).
45. Matiević, E., Espenschied, W. and Kerker, M., *J. Colloid Sci.* **18**, 91 (1963).
46. Espenschied, W., Matiević, E. and Kerker, M., *J. Phys. Chem.* **68**, 2831 (1964).
47. Young, O. and Morrison, J., *J. Sci. Instrum.* **31**, 90 (1954).
48. Thompson, F., Rose, G. and Morrison, J., *J. Sci. Instrum.* **32**, 324 (1955).
49. Craig, A. and McIntosh, B., *Can. J. Chem.* **30**, 448 (1952).
50. Thomann, H., *Physics Fluids*, **9**, 897 (1966).
51. Thomann, H., *Medd. Flygtek. FörsAnst.* **101**, 80 (1964).

52. NOGREN, C., *AIAA Progress in Astronautical Aeronautics*, Academic Press, New York, vol. 9, p. 407 (1963).
53. COX, A., *J. Spacecraft Rockets* **4**, 86 (1967).
54. KARIORIS, F. and FISH, B., *J. Colloid Sci.* **17**, 155 (1962).
55. KARIORIS, F. and WOYCI, C., *Adv. X-ray Analysis*, **7**, 240 (1964).
56. WERLE, D., Armour Research Foundation Contract AF-19, Rept. (1955).
57. WINKEL, G. and JANDER, J., *Kolloid. Zh.* **63**, 5 (1933).
58. JANDER, J. and WINKEL, G., *Kolloid. Zh.* **65**, 292 (1935).
59. CAWOOD, W. and WHYTHLAW-GREY, R., *Trans. Faraday Soc.* **32**, 1048 (1936).
60. PETRYANOV, I., TIKHOMIROV, M. and TUNITSKII, N., *Zh. fiz. Khim.* **15**, 841 (1941).
61. WARBURG, G. and NEGELEIN, E., *Biochem. Z.* **204**, 495 (1929).
62. BARZYNSKI, N. and HUMMEL, D., *Z. phys. Chem., Frankfurt*, **38**, 103 (1963).
63. KOGAN, YA. I., *Dokl. Akad. Nauk SSSR*, **161**, 386 (1965).
64. GAILLAT, R. and CUER, J., *Bull. Soc. chim. Fr.* No. 17, 152 (1959).
65. MAZLOVSKII, A. and FUCHS, N. A., *Kolloid. Zh.* **29**, 127 (1967).
66. CUER, J., ELSTON, J. and TEICHNER, S., *Bull. Soc. chim. Fr.* No. 1, 81 (1961).
67. ZAKUTINSKII, V. and BLYAKHER, I., *Trudy̆ Ural. Khim. Inst.* No. 14, 121 (1967).
68. THOMAS, A., *Combust. Flame*, **6**, 46 (1962).
69. AMELIN, A., *Kolloid. Zh.* **29**, 16 (1967).
70. BRICARD, J., BILLARD, F. and MADELAINE, G., *J. Geophys. Res.* **73**, 4487 (1968).
71. FUCHS, N. A. and OSHMAN, N., *Acta phys. chim. URSS*, **3**, 61 (1935).
72. ORR, C., *et al.*, *J. Meteorol.* **15**, 240 (1958).
73. MATTESON, M. and STOBER, W., *J. Colloid Interface Sci.* **23**, 203 (1967).
74. LIU, B., WHITBY, K. and YU, H., *J. Rech. Atm.* **2**, 396 (1966).
75. STERN, S., *J. Appl. Phys.* **30**, 952 (1959).
76. COX, A., *Astronautics*, **1**, 56 (1962).
77. DROZIN, V., *J. Colloid Sci.* **10**, 158 (1955).
78. MASON, E., *Contempor. Phys.* **4**, 27 (1962).
79. NOLAN, P. and POLLAK, L., *Proc. Roy. Irish Acad. Sci.* **51A**, 9 (1946).
80. POLLAK, L., *Geofis. Pura Appl.* **21**, 75 (1952).
81. POLLAK, L. and DALY, J., *Geofis. Pura Appl.* **36**, 27 (1957).
82. POLLAK, L. and DALY, J., *Geofis. Pura Appl.* **41**, 211 (1958).
83. POLLAK, L. and MURPHY, T., *Geofis. Pura Appl.* **25**, 44 (1953).
84. POLLAK, L. and O'CONNOR, T., *Geofis. Pura Appl.* **39**, 321 (1955).
85. NOLAN, P. and SCOTT, J., *Proc. Roy. Irish Acad. Sci.* **64**, 37 (1964).
86. VERZAR, F., *Geofis. Pura Appl.* **29**, 192 (1954).
87. RICH, T., *Geofis. Pura Appl.* **50**, 46 (1961).
88. POLLAK, L. and METNIEKS, A., *Geofis. Pura Appl.* **47**, 123 (1960).
89. SKALA, G., *Analyt. Chem.* **35**, 702 (1965).
90. VONNEGUT, B., *Proceedings 1st Symposium on Air Pollution, Los Angeles, 1949.*
91. HULL, W. and MÜHLEISEN, R., *Naturwiss.* **41**, 301 (1954).
92. TWOMEY, S., *Geofis. Pura Appl.* **29**, 192 (1954).
93. WIELAND, W., *Z. angew. Math. Mech.* **7**, 428 (1956).
94. LA MER, V., INN, E. and WILSON, J., *J. Colloid Interface Sci.* **5**, 471 (1950).
95. LA MER, V., US Atomic Energy Commission Rept. NYO-512 (1952).
96. VLASENKO, G. and DERJAGUIN, B., *et al.*, *Studies on Clouds, Atmospheric Precipitation and Thunderstorm Electricity*, Moscow, p. 185 (1957).
97. KOGAN, YA. I. and BURNASHEVA, A., *Zh. fiz. Khim.* **39**, 2630 (1959).
98. FUCHS, N. A. and SUTUGIN, A. G., *Kolloid. Zh.* **28**, 131 (1966).
99. AITKEN, J., *Proc. Roy. Soc.* A, **37**, 215 (1916).
100. NOLAN, P. and POLLAK, L., *Geofis. Pura Appl.* **34**, 60 (1955).
101. YAFFE, Y. and CADLE, R., *J. Phys. Chem.* **62**, 510 (1958).
102. TWOMEY, S., *J. Rech. Atmos.* **1**, 81 (1961).
103. ORR, C. and WILSON, T., *J. Colloid Sci.* **19**, 510 (1964).
104. TWOMEY, S. and SEVERINCE, G., *J. Atmos. Sci.* **20**, 392 (1963).
105. BUYKOV, M. and BAKHANOV, V., *Kolloid. Zh.* **28**, 164 (1966).
106. BUYKOV, M., *Kolloid. Zh.* **28**, 636 (1966).
107. SIKSNA, R., *Geofis. Pura Appl.* **50**, 23 (1961).
108. VONNEGUT, B. and NEUBAUER, R., *Bull. Am. Meteorol. Soc.* **34**, 163 (1953).
109. SOUDAIN, G., *J. scient. Mét.* **3**, 137 (1951).
110. BINEK, M., *Rev. trimest. Ass. Prév. Pollution atmos.* **5**, 91 (1963).
111. LANGEVIN, P., *J. de Phys.* **4**, 322 (1965).
112. LISSOVSKI, P., *Acta phys. chim. URSS*, **13**, 157 (1940).

113. ISRAEL, H., *Quart. J. Roy. Meteorol. Soc.* **35**, 341 (1932).
114. MCGREEVY, G., *Arch. Meteorol. Geoph. Bioklimat.* **A14**, 318 (1964).
115. KEEFE, D., NOLAN, P. and SCOTT, J., *Proc. Roy. Irish Acad.* **66A**, 17 (1968).
116. WRIGHT, H., *Proc. Phys. Soc.* **48**, 675 (1936).
117. RICH, T., *Int. J. Air Water Pollution*, **1**, 288 (1959).
118. RICH, T., METNIEKS, A. and POLLAK, L., *Geofis. Pura Appl.* **44**, 233 (1959).
119. FUCHS, N. A., *Geofis. Pura Appl.* **56**, 185 (1963).
120. NOLAN, P. and GUERRINI, V., *Proc. Roy. Irish Acad. Sci.* **43**, 5 (1935).
121. RADUSHKEVICH, L., *Acta phys.-chim. URSS*, **11**, 265 (1939).
122. GORMLEY, P. and KENNEDY, M., *Proc. Roy. Irish Acad. Sci.* **52A**, 166 (1949).
123. GORMLEY, P. and KENNEDY, M., *Proc. Roy. Irish Acad. Sci.* **45A**, 59 (1938).
124. DE MARCUS, W. and THOMAS, J., US Atomic Energy Commission Rept. ORN1-1413 (1952).
125. TWOMEY, S., *J. Franklin Inst.* **275**, 1969 (1963).
126. RODEBUSH, W., US Atomic Energy Commission Rept. OSRD-2050 (1943).
127. POLLAK, L. and METNIEKS, A., *Geofis. Pura Appl.* **37**, 183 (1957).
128. POLLAK, L. and METNIEKS, A., *Geofis. Pura Appl.* **44**, 223 (1959).
129. POLLAK, L., DALY, J. and METNIEKS, A., *Geofis. Pura Appl.* **45**, 249 (1960).
130. NOLAN, P. and SCOTT, J., *Proc. Roy. Irish Acad. Sci.* **65A**, 39 (1965).
131. FUCHS, N. A., STECHKINA, I. and STAROSSELSKI, V., *Brit. J. Appl. Phys.* **13**, 281 (1962).
132. TWOMEY, S., *J. Franklin Inst.* **275**, 163 (1963).
133. METNIEKS, A., *Geofis. Pura Appl.* **61**, 183 (1965).
134. THOMAS, J., *J. Colloid Sci.* **10**, 46 (1955).
135. THOMAS, J., *J. Colloid Sci.* **11**, 107 (1956).
136. TWOMEY, S. and SEVERINCE, C., *J. Atmos. Sci.* **20**, 392 (1963).
137. KOGAN, YA. and REPINA, R., *Dokl. Akad. Nauk SSSR*, **111**, 85 (1956).
138. BRICARD, J., *et al.*, *Staub-Reenthalt. Luft*, **24**, 345 (1964).
139. CHAMBERLAINE, A. and MEGAW, W., *Geofis. Pura Appl.* **36**, 273 (1957).
140. KORPUSSOV, V., KIRICHENKO, V. and OGORODNIKOV, B., *Atomnaya Energiya*, **17**, 221 (1964).
141. BEREJNOJ, V., *Atomnaya Energiya*, **18**, 342 (1965).
142. FUCHS, N. A., *Mechanics of Aerosols*, Pergamon Press, p. 209 (1964).
143. ANDERSON, F., NOLAN, P. and O'CONNOR, T., *Proc. Roy. Irish Acad. Sci.* **66a**, 69 (1968).
144. MEGAW, W. and WIFFEN, K., *J. Rech. Atmos.* **1**, 173 (1963).
145. NOLAN, P. and KUFFEL, P., *J. Atmos. Terrestrial Phys.* **3**, 181 (1953).
146. FUCHS, N. A. and STECHKINA, I., *Ann. Occup. Hyg.* **6**, 27 (1963).
147. KIRSCH, A. and FUCHS, N. A., *Ann. Occup. Hyg.* **11**, 299 (1968).
148. RICH, T., *J. Rech. Atmos.* **2**, 79 (1966).
149. POLLAK, L., O'CONNOR, T. and METNIEKS, A., *Geofis. Pura Appl.* **34**, 177 (1956).
150. POLLAK, L. and MURPHY, T., *Geofis. Pura Appl.* **31**, 66 (1955).
151. LABEIRIE, J., *Brit. J. Appl. Phys.* Suppl. No. 3, Discussion (1954).
152. FUCHS, N. A. and JANKOVKII, S., *Kolloid. Zh.* **21**, 133 (1959).
153. ORR, C. and WILSON, T., *J. Colloid Sci.* **19**, 571 (1964).
154. ORR, C. and MARTIN, R., *Rev. Scient. Instrum.* **29**, 129 (1958).
155. LANGER, G., PIERRARD, J. and YAMATE, A., *Int. J. Air Water Pollution*, **8**, 167 (1964).
156. BAKER, R. (ed.), *Advances in Optical and Electron Microscopy*, Academic Press, New York (1966).
157. MEGAW, W. and WIFFEN, R., *Int. J. Air Water Pollution*, **7**, 501 (1963).
158. CADLE, R., *Particle Size Determination*, Interscience, New York (1955).
159. KALMUS, E., *J. Appl. Phys.* **25**, 87 (1954).
160. BIGG, E., MILES, G. and HEFFERMANN, K., *J. Meteorol.* **18**, 804 (1961).
161. MOSSOP, S. and THORNDIKE, H., *J. Appl. Math. Phys.* **5**, 474 (1966).
162. CARTWRIGHT, J., *Quart. J. Roy. Meteorol. Soc.* **82**, 82 (1956).
163. STECHKINA, I. B., *Proceedings 8th Intercolloid Conference on Stability and Dynamics of Disperse Systems*, Odessa, 1968, p. 51.
164. LECHTMAKHER, S. and RUSER, L., *ibid.*, p. 39.
165. LORD, R., and HARBOUR, A., *AIAA Jl.* **6**, 2 (1968).
166. FUCHS, N. A., *Evaporation and Droplet Growth in Gaseous Media*, Pergamon Press (1961).
167. JER RU MAA, *Ind. Engng. Chem. Fund.* **6**, 504 (1967).
168. SCHÄFER, K., RATING, W. and ENSKOG, A., *Ann. d. Phys.* **42**, 176 (1942).
169. FRISCH, H. and COLLINS, F., *J. Chem. Phys.* **20**, 797 (1952).
170. TIMOFEEV, M. and SCHWETZ, M., *Meteorologia Hidrol.* No. 2 9 (1946).
171. SHERMAN, P., *Rarefied Gas Dynamics, 3rd Symposium*, Academic Press, New York, vol. 2, p. 228 (1963).
172. WELANDER, P., *Ark. Fys.* **7**, 507 (1954).
173. BHATNAGAR, P., GROSS, E. and KROOK, H., *Phys. Rev.* **94**, 511 (1954).
174. BROCK, J., *J. Colloid Interface Sci.* **22**, 513 (1966).

175. Brock, J., *J. Colloid Interface Sci.* **24**, 344 (1967).
176. Jeans, J., *Dynamic Theory of Gases*, 4th ed., Dover, ch. 13 (1925).
177. Davison, B., *Neutron Transfer Theory*, Cambridge (1957).
178. Sahni, D., *J. Nucl. Energy*, **20**, 915 (1966).
179. Smirnov, V., *Trudӯ tsent. aéro. observ. Moscow*, No. 55 (1964).
180. Bradley, P., *et al.*, *Proc. Roy. Soc.* A, **186**, 363 (1948); **193**, 226 (1949).
181. Monchik, L. and Reiss, H., *J. Chem. Phys.* **22**, 831 (1954).
182. Brock, J., *J. Colloid Interface Sci.* **24**, 344 (1967).
183. Brock, J., *J. Phys. Chem.* **68**, 2857 (1964).
184. Smirnov, V., *Trudӯ tsent. aéro. observ. Moscow*, No. 92 (1968).
185. Sang-Wook, K., *AIAA Jl.* **5**, 1208 (1967).
186. Merritt, C. and Weatherstone, R., *AIAA Jl.* **5**, 190 (1967).
187. Okuyama, M. and Zung, J., *J. Chem. Phys.* **46**, 2064 (1967).
188. Horn, K., Bevc, V. and Kaplan, M., *AIAA Jl.* **5**, 721 (1967).
189. Chamberlaine, A., *et al.*, *Discuss. Faraday Soc.* **30**, 162 (1960).
190. Briggs, G. H., *Phil. Mag.* **50**, 630 (1928).
191. Millikan, R., *Phys. Rev.* **21**, 217 (1923); **22**, 1 (1923).
192. Epstein, P., *Phys. Rev.* **23**, 710 (1924).
193. Fuchs, N. A. and Stechkina, I., *Zh. tekh. Fiz.* **33**, 132 (1963).
194. Liu Vi Chang, Pang Sing Chen and Jew, H., *Phys. Fluids*, **8**, 788 (1968).
195. Willis, D., *Phys. Fluids*, **9**, 2522 (1966).
196. Cercignani, C. and Pagani, C., *Phys. Fluids*, **9**, 1167 (1966).
197. Cercignani, C. and Pagani, C., *Phys. Fluids*, **11**, 1395, 1399 (1968).
198. Brock, J., *J. Phys. Chem.* **68**, 2863 (1964).
199. Kennard, E., *Kinetic Theory of Gases*, McGraw-Hill, p. 313 (1938).
200. Springer, G. and Tsai, S., *Phys. Fluids*, **8**, 1361 (1965).
201. Lees, L., *J. Soc. Ind. Appl. Math.* **13**, 278 (1965).
202. Brock, J., *Phys. Fluids*, **9**, 1601 (1966).
203. Cercignani, C. and Pagani, C., *Rarefied Gas Dynamics, 5th Symposium*, Academic Press, New York, vol. 1, p. 555 (1967).
204. Takao, H., *Rarefied Gas Dynamics, 3rd Symposium*, Academic Press, New York, vol. 2, p. 102 (1963).
205. Harbour, P., Phil.D. diss., Cambridge (1963).
206. Junge, C., *Air Chemistry and Radioactivity*, Academic Press, New York (1963).
207. Gentry, G. and Brock, J., *J. Chem. Phys.* **47**, 64 (1967).
208. Liu, B., Whitby, K. and Yu, H., *J. Colloid Interface Sci.* **23**, 367 (1967).
209. Keefe, D., Nolan, P. and Rich, T., *Proc. Roy. Irish Acad.* **62A**, 8 (1962).
210. Keefe, D., Nolan, P. and Rich, T., *Proc. Roy. Irish Acad.* **60A**, 27 (1959).
211. Keefe, D., Nolan, P. and Scott, J., *Proc. Roy. Irish Acad.* **66A**, 17 (1968).
212. Bricard, J., *Problems of Atmospheric and Space Electricity*, Elsevier, Amsterdam, p. 82 (1965).
213. Siksna, R., *J. Rech. Atmos.* **1**, 137 (1963).
214. Fuchs, N. A., *Geofis. Pura Appl.* **56**, 185 (1963).
215. Natanson, G., *Zh. tekh. Fiz.* **30**, 573 (1960).
216. Flanagan, V., *Geofis. Pura Appl.* **64**, 197 (1966).
217. Liu, B., Whitby, K. and Yu, H., *J. Appl. Phys.* **38**, 1592 (1967).
218. Pollak, L. and Metnieks, A., *Geofis. Pura Appl.* **51**, 225 (1962).
219. Einstein, A., *Z. Phys.* **27**, 1 (1924).
220. Cawood, W., *Trans. Faraday Soc.* **32**, 1068 (1936).
221. Clusius, K., *Z. VDI Verfahrenstechnik*, No. 2, 23 (1941).
222. Waldmann, L., *Z. Naturforsch.* **14a**, 589 (1959).
223. Derjaguin, B. and Bakanov, S., *Kolloid. Zh.* **21**, 377 (1959).
224. Mason, E. and Chapman, S., *J. Chem. Phys.* **36**, 627 (1963).
225. Monchik, L., Yu, K. and Mason, E., *J. Chem. Phys.* **39**, 654 (1963).
226. Wang Chang, C. and Uhlenbeck, G., *Engng. Res. Inst. Univ. Michigan*, Report (1951).
227. Brock, J., *J. Colloid Interface Sci.* **23**, 448 (1967).
228. Schmitt, K., *Z. Naturforsch.* **14a**, 870 (1959).
229. Brock, J., *J. Colloid Interface Sci.* **25**, 392 (1967).
230. Dwyer, H., *Phys. Fluids*, **10**, 975 (1967).
231. Astakhov, A. V., *Dokl. Akad. Nauk SSSR*, **161**, 1114 (1965).
232. Brock, J., *Phys. Fluids*, **11**, 922 (1968).
233. Gardner, G., *Chem. Engng. Sci.* **23**, 29 (1968).
234. Schadt, G. and Cadle, R., *J. Colloid Sci.* **12**, 356 (1957).
235. Waldmann, L., *Rarefied Gas Dynamics* (Talbot, ed.), Academic Press, New York, p. 323 (1960).
236. Derjaguin, B. and Yalamov, Yu., *J. Colloid Interface Sci.* **22**, 117 (1966).

237. Brock, J., *J. Colloid Interface Sci.* **27**, 95 (1968).
238. Waldmann, L. and Schmitt, K., *Z. Naturforsch.* **15a**, 843 (1960).
239. Brock, J., *J. Phys. Chem.* **72**, 747 (1968).
240. Rosenblatt, P. and La Mer, V., *Phys. Rev.* **70**, 335 (1946).
241. Jacobsen, S. and Brock, J., *J. Colloid Sci.* **20**, 544 (1965).
242. Brock, J. and Hidy, G., *J. Appl. Phys.* **36**, 1857 (1965).
243. Hidy, G. and Brock, J., *J. Colloid Sci.* **20**, 477 (1965).
244. Friedlander, S. and Wang, C., *J. Colloid Interface Sci.* **24**, 170 (1967).
245. O'Connor, T., *Geofis. Pura Appl.* **31**, 107 (1955).
246. Quon, J., *Int. J. Air Water Pollution*, **8**, 355 (1964).
247. Mayer, J., *J. Chem. Phys.* **1,** 270 (1933).
248. Sutugin, A. G., *Kolloid. Zh.* **29**, 852 (1967).
249. Zebel, G., in *Aerosol Science* (Davies, ed.), Academic Press, London (1966).
250. La Mer, V. and Gruen, R., *Trans. Faraday Soc.* **48**, 410 (1952).
251. Blackman, M., Lisgarten, N. and Skinner, L., *Nature*, **217**, 1245 (1968).
252. Tolman, R., *J. Chem. Phys.* **17**, 33 (1949).
253. Buff, F., *J. Chem. Phys.* **19**, 159 (1951).
254. Buff, F., *Discuss. Faraday Soc.* **30,** 52 (1960).
255. Hill, T., *J. Chem. Phys.* **56**, 527 (1952).
256. Kirkwood, B. and Buff, F., *J. Chem. Phys.* **10**, 991 (1950).
257. Plesner, I., *J. Chem. Phys.* **40**, 1510 (1964).
258. Shcherbakov, L., *Kolloid. Zh.* **14**, 379 (1952).
259. Shcherbakov, L., *Kolloid. Zh.* **23**, 338 (1961).
260. Bellemans, A., *Physica*, **29**, 548 (1963).
261. Cahn, J. and Hilliard, J., *J. Chem. Phys.* **28**, 548 (1958).
262. Cahn, J., *J. Chem. Phys.* **31**, 688 (1959).
263. Hart, E., *Phys. Rev.* **113**, 412 (1959).
264. Hart, E., *Fundamental Phenomena in Material Science*, Plenum Press, New York, p. 37 (1966).
265. Tamman, L., *Z. anorg. allg. Chem.* **110**, 166 (1928).
266. Higuchi, J., *Science Report Tohoku Univ.*, 1st ser., **33**, 237 (1949).
267. Okuyama, M. and Zung, J., *J. Chem. Phys.* **46**, 1580 (1967).

DYNAMICS OF CHARGED SUSPENSIONS

S. L. Soo

Professor of Mechanical Engineering
University of Illinois, Urbana, Illinois 61801

Abstract. Particles or droplets in a gaseous suspension are almost invariably electrically charged. Charging and charge transfer may occur by surface contact or by collection of ions. All particle–fluid interactions, particle–surface interactions and particle–particle interactions are influenced by the electrical effects. The method of continuum mechanics of suspensions permits accurate calculation and correlation of a class of phenomena of diffusion under field forces produced by electrical, gravitational, and hydrodynamic effects. They include coalescence, electric field, and electrification of rain clouds; diffusion and fall-out in the atmosphere, turbulence and dispersion; interaction with a boundary and sampling; and electrical means of collecting particles.

CONTENTS

NOMENCLATURE

a	Radius of a particle, or a radius.
a_a	Molecular radius of air.
a_e	Radius of a sphere of equivalent volume.
a_m	Mean particle radius.
A	Area.
b	A distance or length.
$B(S)$	A function of speed ratio S.
c	Specific heat at constant pressure, or a length.
c_p	Specific heat of a particle.
c_v	Specific heat at constant volume.
$c^{(q)}$	Mole-fraction of component (q).
C	Cunningham correction factor, or a capacitance.
C_D	Drag coefficient of a particle.
C_{Dfm}	Drag coefficient in free molecule flow.
C_f	Skin friction coefficient.
C_L	Lift coefficient.
d	Characteristic length for charge transfer.
D	Diffusivity.
e	Electronic charge (1.6×10^{-19} coulomb).
E	Modulus of elasticity, electric field, or energy.
$\mathbf{E}$	Vectorial electric field.
f	Distribution function.
$\boldsymbol{f}$	Force acting on a particle.
$\boldsymbol{f}_D$	Drag force acting on a particle.
$\boldsymbol{f}_e$	Force due to electrostatic repulsion.
$\boldsymbol{f}_L$	Lift force.
$\boldsymbol{f}_q$	Electric drag force.
$\mathbf{f}$	Force vector.
$\mathbf{f}_e$	Force vector due to external field.
F	Time constant for momentum transfer from fluid to particle.
$F^{(qp)}$	Time constant for mutual interaction of particles.
F'	A drag constant defined after eqn. (4.33).
F^*	A correction coefficient for non-Stokesian motion.
$\boldsymbol{F}$	Force per unit mass.
g	Gravitational acceleration.
$\boldsymbol{g}$	Rate of heat generated per particle, including radiation input.
G	Time constant for heat transfer between fluid and particle.
$G^{(qp)}$	Time constant for heat transfer between particle clouds.
G_1^*, G_2^*	Energy function defined by eqns. (3.29) and (3.30).

G	Rate of heat generated per unit volume, including radiation.
h	Heat transfer coefficient or charge transfer coefficient.
H	Inverse relaxation time for charge transfer.
i	Electric current.
J	Flux of particles or current density.
J_q	Heat flux.
k	Boltzmann constant.
k_1	$(1-\nu_1^2)/\pi E_1$.
k_u	Factor accounting for interfacial velocity.
K	Mobility, a constant, temperature coefficient, or coagulation coefficient.
K_p	Mobility of a particle.
K_m	Effectiveness of momentum transfer from particle to fluid.
L_p	Particle–fluid interaction length for momentum.
L_T	Particle–fluid interaction length for energy.
m	Mass of a particle.
m_a	Average molecular mass of air.
m_z, m_i	Mass of an ion.
$\dot{m}$	Total flow rate.
$\dot{m}_{pw}$	Collision rate at the wall.
m^*	Mass ratio of particulate to fluid phase.
$\dot{m}^*$	Mass flow ratio of particulate to fluid phase.
M	Molecular weight or electric moment.
n	Number density.
n_p	Number density of particles.
n_e	Free electron density.
n_z, n_i	Ion density.
n_Z	Number density of particles with Z charges.
N	Number of ions.
N_D	Diffusion convection number.
N_{DF}	Diffusion response number.
N_{DV}	Drift-diffusion parameter.
N_{es}	Electrosurface number.
N_{et}	Electrothermal number.
$(N_{et})_E$	Electrothermal number based on external field.
$(N_{et})_p$	Electrothermal number based on particle charge.
N_{ev}	Electroviscous number, or electrodiffusion number.
N_{Fr}	Froude number.
N_{Im}	Impact number.
N_{Kn}	Knudsen number.
N_{Kp}	Knudsen number for particle–fluid interaction.
N_m	Fluid–particle momentum number.
$(N_m)_g$	Fluid–particle momentum transfer number in gravitational field.
N_M	Mach number.
N_{Nu}	Nusselt number.
N_{Pe}	Peclet number.
N_{Pr}	Prandtl number.

N_{Re}	Reynolds number.
$(N_{Re})_e$	Electric Reynolds number.
N_S	Shear response number.
N_{Sc}	Schmidt number.
N_{St}	Stanton number.
N_{SV}	Space-charge parameter.
N_{We}	Weber number.
N	Rate of mass transfer in number of moles per unit time.
p	Electric dipole moment.
P	Static pressure.
q	Electric charge.
Q	Charge transfer per collision.
q_1	Specific charge per unit volume.
q_0	Maximum charge for given band structure.
q/m	Charge to mass ratio.
r	Radius, radial coordinate.
r^*	Ratio of the reflected speed to incoming speed of collision.
$\mathbf{r}$	Position vector.
R	Radius or pipe radius.
$\bar{R}$	Gas constant based on mass.
S	Molecular speed ratio.
t	Time.
T	Absolute temperature.
T_w	Wall or surface temperature.
T_∞	Temperature at infinity from particle.
u	Velocity, internal energy.
U	Velocity.
ΔU	Relative speed.
$\langle U_p^2 \rangle^{1/2}$	Intensity of particle motion.
$\mathbf{U}$	Vectorial fluid velocity.
$\mathbf{U}_p$	Vectorial particle velocity.
v	Speed.
$\langle (\Delta u)^2 \rangle^{1/2}$	RMS relative velocity of particle to fluid.
$\bar{v}$	Mean thermal speed of ions.
$\mathbf{v}_p$	Volume of a particle.
V	Electric potential.
$\mathbf{v}$	Volume.
x, y	Cartesian coordinates.
$\mathbf{x}$	Position vector.
x_i	Position coordinate.
Z	Number of electron holes.
Z_s	Saturation charge.
z	Axial coordinate.
α	$(N_{et})_p$, accommodation coefficient, a constant, or Townsend ionization coefficient.
α_1	Collision parameter.

β	Empirical constant or diffusion parameter.
γ	Ratio of specific heat.
Γ	Rate of generation.
δ_{ij}	Kronecker delta.
Δ	Deformation tensor.
$\delta^*, \delta_p^*, \delta_{pg}, \theta, \theta_p, \delta_{p1}$	Momentum integrals of boundary layer calculations.
ε	Permittivity of material.
δ	Ion sheath thickness, or boundary layer thickness.
ε_0	Permittivity of free space.
ζ	Bulk viscosity.
$\alpha, \beta, \eta, \gamma$	Characteristic groups of pipe flow.
η_{co}	Fraction of collisions that leads to coalescence.
η	Fraction impacted, or collection efficiency, or Kolmogoroff length scale of turbulence.
θ	Azimuthal angle, angle of inclination.
Θ	Dilatation tensor.
$\bar{\kappa}$	Thermal conductivity of the fluid material.
κ_m	Thermal conductivity of a suspension.
$\kappa_m^{(q)}$	Thermal conductivity of component (q) in the mixture.
$\bar{\kappa}_p$	Thermal conductivity of material constituting the particulate phase.
λ	Mean free path.
Λ	Wavelength, microscale.
λ_D	Debye shielding distance.
τ	Shear stress, time, dimensionless time, or temperature ratio.
ξ	Parameter Fx/U.
μ_2	$\zeta-(2/3)\mu$.
μ_m	Viscosity of a mixture.
$\mu_m^{(q)}$	Viscosity of component (q) in the mixture.
$\bar{\mu}$	Viscosity of the fluid material.
$\bar{\mu}_p$	Viscosity of the material constituting the particulate phase.
ν	Poisson ratio of the material.
ν_j	Stoichiometric coefficient.
$\bar{\nu}$	Kinematic viscosity of the fluid material.
ρ	Density of fluid phase.
ρ_p	Density of particulate cloud.
$\bar{\rho}$	Material density of fluid phase.
$\bar{\rho}_p$	Density of material.
σ_s	Surface tension.
σ	Electrical conductivity.
Φ	Volume fraction of particulate phase.
φ	Polar angle, electrokinetic potential, or potential of work function.
φ_0	Potential difference of droplets, or surface potential.
ψ	Electric flux.
ψ_p	Stream function of particulate phase.

Superscripts

*	Dimensionless quantities.

(0), (1) ...	Order of perturbation.
(q), (p)	Component (q) or (p) of mixture.

Unsubscripted quantities	Fluid phase.

Subscripts

a	Molecules
c	Contact.
e	Electron.
i	Ion.
i, j, k	ith, jth, or kth component.
m	Mixture.
o	Initial conditions or reference conditions.
p	Particulate phase/particle.
pr	Reflected particle.
pi	Incoming particle.
w	At wall.

INTRODUCTION

VARIOUS hydrodynamic phenomena of transport, settling, coalescence, and collection of particulate matter in gaseous suspension are strongly influenced by the fact that they carry electric charges by nature or by design. Such influences range from being a perturbation on the hydrodynamic behavior of an aerosol suspension to being an overriding effect. That aerosol particles are seldomly electrically neutral is substantiated by the following well-known facts:

Deposition of dust on ceilings.
Collection of dust particles smaller than filter openings.
Residual collection efficiency of an electrostatic precipitator at power-off.

The following chapters delineate recent results on the dynamics of dilute gaseous suspensions of charged aerosols pertinent to the atmospheric sciences. For general fundamentals of hydrodynamics and electrohydrodynamics of multiphase systems, the readers are referred to a recent monograph.(1) The present review emphasizes aspects not included in that volume and supersedes some of the earlier results, although this review is self-contained.

While many of the basic relations are applicable to liquid suspensions, our concern in this part is mainly with gaseous suspensions of solid particles and liquid droplets.

In preparation for treating dynamics of charged suspensions, we shall first consider charging and transport properties. This is followed by general formulation and examples of various electrohydrodynamic systems.

CHAPTER 1

SOURCES OF ELECTRIC CHARGES

Charging of particulate matter such as dusts, lints, and droplets occurs by surface contact, collection of atmospheric electricity, or from a corona discharge by kinetic, electrodynamic, and aerodynamic capture of ions, and by thermionic and photoemission.

1.1. Charging by surface contact

If an aerosol makes contact with any surface at all, charging occurs due to the difference in Fermi levels between contacting surfaces. Surface contact provides equalization of the Fermi levels of the surfaces by charge transfer resulting in a contact potential. Subsequent separation leaves the one with a higher Fermi level of the two surfaces positively charged, and the other becomes negatively charged. Charges on sandstorms over deserts may be attributed to this surface effect as well as to atmospheric electricity.(2) (See next section.) Charging by surface contact also occurs due to drop-breaking and collision.(3) In spite of the fact that man's knowledge of electricity began with charging by surface interaction (tribo-electric effect),(4) quantitative determination of the magnitude of electric charge by contact and impact has been made only recently.(5)

The basic relation for charge transfer by impact or collision is seen by considering the case of collision of two spheres of solid materials denoted as 1 and 2, of radii a_1 and a_2, masses m_1 and m_2, densities $\bar{\rho}_1$ and $\bar{\rho}_2$, conductivities σ_1 and σ_2 of their materials, potentials V_1 and V_2, velocities u_1 and u_2 with directions making an angle θ. With such a highly idealized system, the average conditions during impact are given by the Hertzian theory on collision mechanism of two elastic spheres. The duration of impact Δt_{21} and area of contact A_{21} are given as:(1)

$$A_{21} = \frac{\pi a_1 a_2}{a_1+a_2}\alpha_1 \tag{1.1}$$

and

$$\Delta t_{21} = \frac{2.94}{\Delta u_{21}\cos\theta}\alpha_1, \tag{1.2}$$

where $\Delta u_{21} = |\mathbf{u}_2-\mathbf{u}_1|$ is the velocity of approach at the beginning of impact. α_1 is the distance between the two spheres approaching each other at the instant of maximum compression. With reduced mass consideration, we have

$$\alpha_1 = \left[\frac{15}{16}\pi\left(\frac{1+r^*}{2}\right)(\Delta u_{21})^2\cos\theta\left(\frac{m_1 m_2}{m_1+m_2}\right)\left(\frac{a_1+a_2}{a_1 a_2}\right)^{1/2}(k_1+k_2)\right]^{2/5}, \tag{1.3}$$

where k's are the elastic properties which are defined as $k = (1-\nu^2)/\pi E$, ν and E are the

Poisson ratio and the modulus of elasticity of each of the materials respectively. r^* is the ratio of rebound speed to incoming speed.

The product $A_{21}\Delta t_{21}$ is now

$$A_{21}\Delta t_{21} = \frac{2.94\pi}{\Delta u_{21}\cos\theta}\frac{\pi a_1 a_2}{a_1+a_2}\alpha_1^2$$

$$= 0.97\pi a_2^3\left(1+\frac{m_2}{m_1}\right)^{-4/5}\left(1+\frac{k_2}{k_1}\right)^{4/5}\left(1+\frac{a_2}{a_1}\right)^{-3/5}\left(\frac{k_1}{k_2}\right)^{2/5}\left(\frac{\bar{\rho}_2}{\bar{\rho}_1}\right)^{2/5}\times$$

$$\times\left[(N_{Im})_{21}\right]^{4/5}\frac{(\cos\theta)^{-1/5}}{\Delta u_{21}} \equiv (A_{21}\Delta t)_0(\cos\theta)^{-1/5}, \quad (1.4)$$

where the dimensionless impact number is defined as

$$(N_{Im})_{21} = 5\pi^2(\Delta u_{21})^2[\bar{\rho}_1\bar{\rho}_2]^{1/2}(k_1k_2)^{1/2}\left(\frac{1+r^*}{2}\right) \quad (1.5)$$

consisting of elastic properties of particle materials only.

The current density through the contact area A_{21} is given by

$$J_{21} = h_{21}(V_2-V_1) = (\sigma_1/d_1')(V_c-V_1) = (\sigma_2/d_2')(V_2-V_c), \quad (1.6)$$

where h_{21} is the charge transfer coefficient through the contact surfaces, V_c is the potential at the contact area, d_1' and d_2' are characteristic lengths of charge transfer in each particle. d_1' and d_2' are not known in detail, but are expected to be each of the order of a_1, a_2 respectively. The time rate of change of potential for each particle is

$$C_1\frac{dV_1}{dt} = -C_2\frac{dV_2}{dt} = J_{21}A_{21} = A_{21}h_{21}(V_2-V_1) \quad (1.7)$$

after eliminating V_c, where C_1 and C_2 are the capacitances of the particles respectively, $C = 4\pi\varepsilon a$, ε is the permittivity, and

$$h_{21} = \left[\frac{\sigma_1\sigma_2}{d_1'd_2'}\right]^{1/2}\left[\sqrt{\left(\frac{\sigma_2 d_1'}{\sigma_1 d_2'}\right)}+\sqrt{\left(\frac{\sigma_1 d_2'}{\sigma_2 d_1'}\right)}\right]^{-1}. \quad (1.8)$$

For free particles, having charges q_1 and q_2, and work functions φ_1 and φ_2

$$V_2-V_1 = \left(\varphi_2+\frac{q_2}{4\pi\varepsilon_2 a_2}\right)-\left(\varphi_1+\frac{q_1}{4\pi\varepsilon_1 a_1}\right). \quad (1.9)$$

For this case, integration of eqn. (1.7) for variables q's gives the net charge transfer per impact:

$$Q_{21} = \frac{C_1C_2}{C_1+C_2}(V_2-V_1)(1-e^{-\alpha\Delta t}) = -Q_{12}, \quad (1.10)$$

where

$$\alpha = A_{21}h_{21}(C_1+C_2)/C_1C_2.$$

The charge transfer coefficient h_{21} given by eqn. (1.8) is based on clean surfaces. In reality, both h_{21} and φ's are influenced by adsorbed surface layer and humidity.[6,7] In the case of mineral particles, the conductivity in logarithm scale increases almost linearly with an increase in relative humidity.[8] For collision of 16μ coal dust on a 0.5 in. steel sphere at 60 to 120 fps, h_{21} ranges from 2 to 5×10^{-8} mho/cm^2, which is much lower than that based on material properties in eqn. (1.8). This magnitude was measured on coal dust having a saturation charge to mass ratio of 10^{-3} coul/kg in the flow system according to

$$\varphi+\frac{q}{4\pi\varepsilon a} = \text{constant} \quad (1.11)$$

by collision with a brass surface (pipe). The average charge transfer per impact vs. the impact number is shown in Fig. 1 for probes of steel spheres of various diameters.

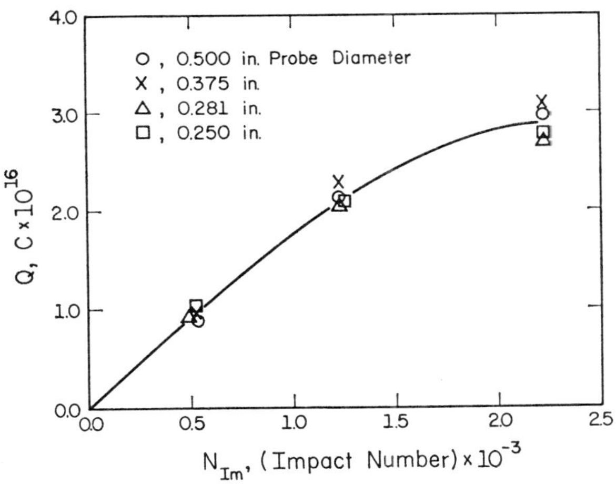

FIG. 1. Average charge transfer per impact and impact number.[5] Measurements made by L. Cheng.

1.2. Atmospheric electricity

Atmospheric electricity has been the subject of intensive studies; recent results are presented in the volumes edited by Thrane[9] and Coroniti.[10] Although ion concentrations in the upper atmosphere from 80 km up, that is above the *E*-layer, are relatively well defined,[11] below that the reported concentration of electrons and ions varies over wide ranges from 40 to 90 km. Below 40 km significant variation occurs with weather and geographical locations. A summary of results from various sources is given in Fig. 2.[3,9,10] It is seen that at all altitudes ions are generated by cosmic radiation with a net space charge at lower atmosphere due to drift of charged particles of different mobility toward the earth. Ionization overland at lowest layers may also occur due to radioactive substances in the earth's crust. The earth's charge also varies due to precipitation and lightning discharges. Takahashi[12] studied the thermoelectric effect in ice and reported activation energy of 2 ev for H_3O^+ and 1 ev for a single ice crystal. It was demonstrated that even a waterfall may have an effect on atmospheric electricity.[13]

1.3. Gaseous discharge

Man-made ionized gases include those produced by corona and arc discharges. Corona discharge at atmospheric conditions has been studied for voltages up to 100 kv. The magnitude of corona current for a 0.3 mm wire in a 20 cm diameter tube is to the order of 1 mA per meter length at 50 kv.[14] An arc discharge, depending on temperature and pressure, gives an extent of ionization from a fraction of a percent to fully ionized conditions.[11,15]

Charging of aerosols by collection of electrons and ions follows identical mechanism from either atmospheric electricity or gaseous discharge,[16] and will be treated in the next chapter.

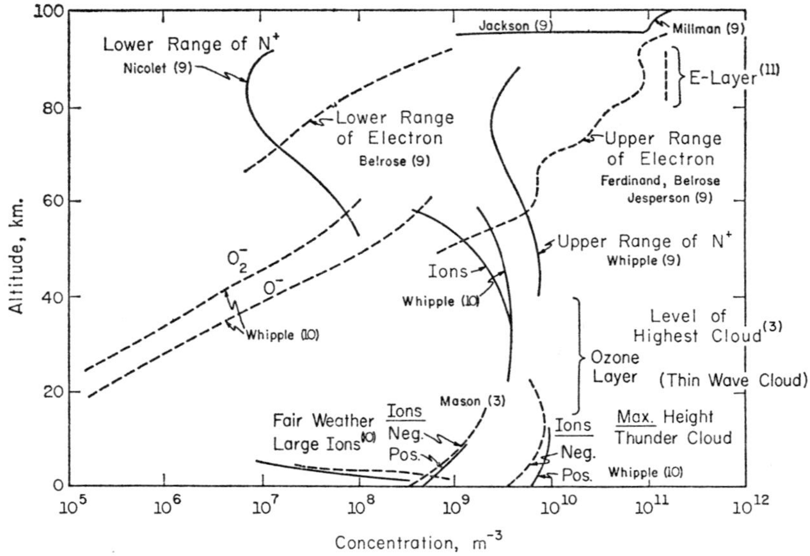

FIG. 2. Magnitude of atmospheric electricity.[3, 9, 10]

1.4. Emission charging

Charging by thermionic emission[1, 6] occurs only at very high temperatures, usually above 1000°K. Charging by photoemission depends on the frequency range of light source and material constituting the aerosol.[17] Water has a threshold frequency corresponding to a wavelength of 1750 Å and that for ice between 1800–2000 Å.[13] Due to significant attenuation of solar radiation occurring in the ozone layer, there is negligible radiation below 2920 Å at the surface of the earth, below 2200 Å at 30 km altitude, 1600 Å at 110 km.[18] Thus photoemission charging of cloud particles is usually insignificant. However, photo-ionization of a gas[18] does occur at low altitudes. The charges may subsequently be collected by aerosol particles.

CHAPTER 2

CHARGING OF PARTICULATE MATTER BY COLLECTION

When a conducting sphere of radius a is placed in a uniform electric field E_0 with an initially uniform unipolar ion density n_0, the potential distribution V is given by the Poisson equation

$$\nabla^2 V = -qn_z/\varepsilon_0, \tag{2.1}$$

where ε_0 is the permittivity,† q is the charge per ion ($q = -e$ for an electron), and n_z is the distributed ion density. The boundary conditions are electric field $E = -\nabla V = E_0$ at infinity and $E = -Ze/4\pi\varepsilon_0 a^2$ for Z charges ($Z > 0$ for positive charges and $Z < 0$ for negative charges) on the surface at a given time. It is readily shown that at any point on the sphere,

$$E_a = 3E_0 \cos\theta - (Ze/4\pi\varepsilon_0 a^2), \tag{2.2}$$

where θ is the azimuthal angle with the direction of E_0 in the spherical coordinate.

2.1. Field charging

For a conducting particle of radius a much larger than the mean free path λ of the ions ($a \geq 1\ \mu$), the random motion of the ions need not be considered and the total electric flux ψ entering the sphere is given by

$$\psi = \int_0^{\pi} 4\pi\varepsilon_0 E_a 2\pi a^2 \sin\theta\, d\theta = 12\pi\varepsilon_0 a^2 E_0[1-(Ze/12\pi\varepsilon_0 E_0 a^2)]^2. \tag{2.3}$$

At saturation, $\psi = 0$, and

$$Z_s = 12\pi\varepsilon_0 E_0 a^2/e. \tag{2.4}$$

The rate of charging is given by the charging current i,

$$i = n_z qK\psi/4\varepsilon_0 = d(Ze)/dt, \tag{2.5}$$

where K is the ion mobility. Integration gives

$$-n_z qKt/4\varepsilon_0 = (Z/Z_s)/[1-Z/Z_s]. \tag{2.6}$$

White[(14)] showed that for $n_z = 5\times 10^{14}\ \text{m}^{-3}$, $K = -2.2$ (cm/sec)/(v/cm); the $q = -e$, the time to reach $Z/Z_s = 1/2$ is 2 msec. When applied to dielectric particles, eqn. (2.4) is modified to

$$Z_s = 12\pi\varepsilon_0 E_0 a^2[1+2(\varepsilon_r-1)(\varepsilon_r+2)^{-1}]/q, \tag{2.7}$$

† We shall use MKS units throughout this part, $\varepsilon_0 = 8.854\times 10^{-12}$ farad/m, or coul²/joule m, or coul/volt m, or sec/ohm m.

where ε_r is the dielectric constant, $\varepsilon_r = \varepsilon/\varepsilon_0$, ε is the permittivity of the material. Schultz and Wiech[19] showed that electrostatic atomization in an intense electric field produces a maximum charge to mass ratio (q/m).

$$(q/m)_{\max} \sim [4\varepsilon_0\sigma_s/\pi f r_m r^2 \bar{\rho}_p^2]^{1/2} \times 10^{-6}\ \text{coul/kg}, \tag{2.8}$$

where σ_s is the surface tension of the liquid and $\bar{\rho}_p$ is its density, r_m is the molecular radius, r is the radius of the meniscus at the instant of dispersion, the function f is given by

$$f = f(\varepsilon_r, \bar{\rho}_p) = (\varepsilon_r - 1)\varepsilon_r^{-1} - \bar{\rho}_p \varepsilon_r^{-2}(d\varepsilon_r/d\bar{\rho}_p). \tag{2.9}$$

Henriquez[20] showed that $f \sim 0.55$ to 1 for non-polar unassociated liquids.

2.2. Charging by diffusion of ions

For radius a of the order of 0.5 μ or less, random motion of ions must be accounted for and the contribution of external electric field becomes less significant.[21–24]

A spherical particle of radius a is introduced into a region of unipolar ions of given density n_{z0} and electric field E_0. It acquires charge due to ions striking the particle. As the charge begins to build up, its repelling force redistributes the ions in the vicinity.

With uniform external field E_0, from solving eqn. (2.1), the region outside the sphere at radius R has a potential[22]

$$V = E_0 \cos\theta[R - (a^3/R^2)(\varepsilon_r - 1)(\varepsilon_r + 2)^{-1}]. \tag{2.10}$$

The charge on the sphere gives a potential $V = Ze/4\pi\varepsilon_0 R$, the potential of the ion at radius R and azimuthal angle θ with the direction of E_0 is

$$V_R = (Ze/4\pi\varepsilon_0 R) + (E_0 a)[(R/a) - (a^2/R^2)(\varepsilon_r - 1)(\varepsilon_r + 2)^{-1}] \cos\theta. \tag{2.11}$$

Denote N_{et} as the electrothermal number; that is, the ratio of electrostatic energy to thermal energy, or

$$\left.\begin{aligned} (N_{et})_p &= Ze^2/4\pi\varepsilon_0 akT \\ \text{and} \qquad & \\ (N_{et})_E &= eE_0 a/kT, \end{aligned}\right\} \tag{2.12}$$

where k is the Boltzmann constant and T is the absolute temperature, subscripts p and E are for the particle and the external field respectively, and we have

$$eV_R/kT = (N_{et})_p(a/R) + (N_{et})_E[(R/a) - (a^2/R^2)(\varepsilon_r - 1)(\varepsilon_r + 2)^{-1}] \cos\theta, \tag{2.13}$$

and at $R = a$,

$$eV_a/kT = (N_{et})_p + (N_{et})_E[1 - (\varepsilon_r - 1)(\varepsilon_r + 2)^{-1}] \cos\theta. \tag{2.14}$$

The local field changes the ion density distribution according to eqn. (2.1). Based on a transformation by Jost,[25] a numerical solution of eqn. (2.1) was given in detail by Murphy.[26]

The case of $E_0 = 0$ is extremely simple when applied to unipolar ions. Taking into account the curvature of the ion free-path, Liu, Whitby, and Yu[27] showed that the particle charge acquired from unipolar ions is given by the White equation[14]

$$(N_{et})_p = \frac{Ze^2}{4\pi\varepsilon akT} = \ln(1 + \tau), \tag{2.15}$$

where $\tau = (n_{z0}\, e^2 a^2/4\varepsilon_0 kT)(vt/a)$, $v = (8kT/\pi m_z)^{1/2}$, the mean thermal speed of ions.

Equation (2.15) was first given by White based on the approximation that every ion striking the particle is captured. From an experimental study of the diffusion charging of mono-dispersed aerosol particles of dioctyl phthalate of 1.35 μ and 0.65 μ in air at 0.0311 to 0.960 atm, Liu, Whitby, and Yu[(28)] further showed that $(N_{et})_p$ is independent of a/λ and the White equation is valid.

2.3. Charging in an ionized gas

The problem of interaction of a spherical particle with a slightly ionized gas was treated by Rosen[(29)] and Dimick and Soo.[(30)] The latter extended the method of Murphy, Adler, and Penney[(24)] for the case of zero external field to interaction with an ionized gas with presence of ions of both signs.

Since the electrons have a higher random velocity than the positive ions by virtue of their small mass, the particle tends to assume a negative charge. For a gas with a low degree of ionization the ion–gas atom and electron–gas atom collisions will dominate over the electron–ion collisions by roughly the ratio of the neutral gas atoms to the positive ion density. An additional assumption is made that there is 100% charge attachment of all electrons and ions arriving at the particle surface.

The interparticle spacing is taken to be larger than an equivalent Debye shielding distance $\lambda_D = (\varepsilon_0 kT/n_{e0} e^2)^{1/2}$ for electron charge density n_{e0}. This limits the particle densities to approximately $n_p = (2\lambda_D)^{-3}$. It was shown that the rate of arrival of electrons and ions at an area dA are given by, for $\alpha = (N_{et})_p$,

$$\left.\begin{aligned} dN_e/dt\big|_{dA} &= (8kT/\pi m_e)^{1/2} n_{e0} \exp(\alpha), \\ dN_i/dt\big|_{dA} &= (8kT/\pi m_i)^{1/2} n_{i0} \exp(-\alpha), \end{aligned}\right\} \tag{2.16}$$

where t is the time, m is the mass, dN/dt is the rate of arrival, subscript e for electrons, i for ions, and

$$n_{e0}/n_{i0} = (m_e/m_i)^{1/2} \exp(2\alpha) \tag{2.17}$$

for $n_{e0}+Zn_p = n_{i0}$. It is further shown by Soo[(1)] that in the absence of an electric field the maximum negative charge that can be attained from an ionized gas corresponds to

$$\alpha > -\tfrac{1}{4} \ln (m_i/m_e). \tag{2.18}$$

It is therefore readily seen that the aerosol particles are usually charged by atmospheric electricity. Examples are charges of 2×10^{-2} coul/kg (2 electrons on 0.25 μ) on tobacco smoke, and 2×10^{-4} coul/kg (720 electrons on 10.5 μ) on natural fogs.[(31)] The values of α are nearly 0.8 and 8 respectively.

Charge to mass ratios (q/m) encountered in atmospheric science range from 10^{-6} to 10^3 coul/kg, and latter by intense field charging in a vacuum.[(19)] Extensive data are presented in the treatise by White.[(14)]

Study by Soo, Wu, and Dimick[(16)] further took into consideration mobility limited diffusion of electrons and ions, and the effect of disparity in electron and ion temperatures in an ionized gas at non-equilibrium.

The particles are assumed to be conductors; even oxides would have substantial conductivity at sufficiently high temperatures. Secondary emission of electrons by impact of ions is negligible for incident ions with energies less than 5 ev. Considerations are given to volume recombination and surface recombination. Reflection of ions and electrons from the surface is negligible due to significant image force.

The equations for the mobility limited diffusion currents of ions and electrons are:[32]

$$-D^{+}\frac{dn_i}{dr}+n_iK^{+}E(r)=J_i(r), \tag{2.19}$$

and

$$-D^{-}\frac{dn_e}{dr}-n_eK^{-}E(r)=J_e(r). \tag{2.20}$$

D^+ and D^- are diffusion coefficients, and K^+ and K^- are the mobilities of positive ions and electrons, with r measured from the center of the particle of radius a. At steady state,

$$J_i(r)=J_e(r)=\frac{a^2}{r^2}J_0. \tag{2.21}$$

At the surface of the particle (subscript p),

$$n_{ea}=n_{ep}\exp\left(-\frac{\varphi_0}{kT_p}\right), \tag{2.22}$$

where n_{ep} is the free electron density in the solid, φ_0 is the work function; this drop in electron density occurs over a thickness of 10^2 Å.

The field around the particle is self-consistent. However, for small ion and electron densities compared to that of neutrals, the radial distribution of the former will be determined by the diffusivities and the field of the particle

$$E=\frac{Ze}{4\pi\varepsilon_0 r^2} \tag{2.23}$$

neglecting the space charge due to the field of the ions. Further,

$$K/D=e/kT. \tag{2.24}$$

Substituting into eqns. (2.19) and (2.20) and integrating gives

$$n_i(r)=\frac{J_0a}{D^{+}\alpha_i}\left[1-\exp(\alpha_i)\left(1-\frac{a}{r}\right)\right], \tag{2.25}$$

$$n_e(r)=\frac{-J_0a}{D^{-}\alpha_e}\left[1-\exp(-\alpha_e)\left(1-\frac{a}{r}\right)\right]+n_{ea}\exp(-\alpha_e)\left(1-\frac{a}{r}\right), \tag{2.26}$$

where

$$\alpha_e=\frac{Ze^2}{4\pi\varepsilon_0 akT_e},$$

$$\alpha_i=\frac{Ze^2}{4\pi\varepsilon_0 akT_i},$$

at large distances $>a$, n_i and n_e are those of the unperturbed plasma.

Eliminating J_0, we get, for $T_e \sim T_i$, $a/r \sim 0$

$$Z=\frac{4\pi\varepsilon_0 akT_i}{e^2}\ln\beta, \tag{2.27}$$

$$J_0=\frac{n_iD^{+}\ln\beta}{a[l-\beta]}. \tag{2.28}$$

At

$$n_{ea}=n_e-\left(\frac{D^{+}}{D^{-}}\right)n_i, \tag{2.29}$$

where

$$\beta = (n_{ea}D^- + n_i D^+)/n_e D^-$$

and $Z = 0$ when the particle is at plasma potential.

Experimental results showed that particulate matter tends to acquire negative charges. Cases are: 50 μ tungsten particles in argon at $T_e = 3500°K$, $T_i = 1250°K$, $Z \sim -10^4$; 30 μ alumina, $T_e = 2800°K$, $T_i = 1000°K$, $Z \sim -5 \times 10^3$.

2.4. Charging of a cloud of particles

The above deals with charging of a single particle for a given local ion concentration. When applied to a cloud of particles or droplets, however, perturbation of local charge density by charge collection occurs. Thus, even a monodispersed cloud of particles will not be uniformly charged. For n_0 uncharged droplets per unit volume, at equal positive and negative conductivity, distribution of charges of different polarity on the drops with charges of the magnitude of Z electronic charges was given by Fuchs[10] as

$$n_Z = n_0 \exp\left[-Z^2 e^2/8\pi\varepsilon_0 akT\right], \tag{2.30}$$

where n_Z is the number density of particles with Z charges.

It was shown empirically that the charge of a droplet is given by

$$q = 4\pi\varepsilon_0 \varphi a, \tag{2.31}$$

where φ is a potential and $q \sim 200\, a$ electronic charges for a in microns, when the surrounding air is neutral, $\varphi = -\varphi_0$, and the electrokinetic potential ~ 0.3 v. q is then the preponderant negative charge that could be acquired by a drop.

In clouds and fogs, acquisition of negative charges by some particles would leave a substantial concentration of positive ions; thus some particles will be positively charged.

For a cloud with particle size distribution following

$$f(a) = dn(a)/da = (5/2)^5 \pi^{-1}(\rho_p/\bar{\rho}_p)(a^2/a_m^6) \exp\left[-5a/a_m\right], \tag{2.32}$$

where a_m is the mean particle size, the specific charge per unit volume of a droplet is given by

$$q_1 = 5\varepsilon_0 \varphi/16\pi a_m^2. \tag{2.33}$$

Gunn,[33] however, suggested that a cloud of particles of various sizes would acquire charges proportional to the square of particle radius. Further detailed study is deemed desirable to resolve this difference.

A general account of electrical behavior of aerosols was given in a survey by Whitby and Liu[34] including charging, collection, and coagulation.

CHAPTER 3

INTERACTIONS BETWEEN FLUID AND AEROSOL PARTICLES

Preparatory to the discussion of dynamics of aerosol suspensions, we now consider various interactions between the fluid and individual aerosol particles. The interactions for the case of uncharged particles suspended in a continuum fluid have been completely explored. A brief review at this stage might offer some examples for further development of the knowledge of interaction parameters of charged systems.

Taking the case of a cloud of identical particles in a suspension, we identify the density of the particle cloud as ρ_p and that of the particle material $\bar{\rho}_p$. The volume fraction Φ of particles in the space occupied by the suspension is then

$$\Phi = \rho_p/\bar{\rho}_p. \tag{3.1}$$

The average interparticle spacing is related to Φ according to, assuming the particles are spheres of radius a,

$$n_p^{-1/3} = (4\pi/3\Phi)^{1/3}a, \tag{3.2}$$

where n_p is the number density of particles. Thus the mean interparticle spacing is nearly 2 particle diameters for $\Phi \sim 0.06$. Defining the density of the fluid phase of the suspension as ρ, its relation to the density of the fluid material $\bar{\rho}$ is

$$\rho = \bar{\rho}(1-\Phi). \tag{3.3}$$

The mass ratio of particle to fluid phase is

$$m^* = \rho_p/\rho = \Phi(1-\Phi)^{-1}\bar{\rho}_p/\bar{\rho}. \tag{3.4}$$

For water droplets in the air, $\bar{\rho}_p/\bar{\rho} \sim 840$. Hence we have $\Phi \sim 0.01$ for $m^* \sim 8$, giving an interparticle spacing of 3.5 diameters. It is noted that interesting cases in atmospheric science rarely have mean interparticle spacing less than 10 diameters.[31] We shall deal with the situation where the assumption of dilute suspension is applicable.

3.1. Fluid forces on solid particles

Comparison with results of Ergun and Orning[35] for solid spheres showed that, for $\Phi < 0.08$, drag relations for single spheres are sufficiently accurate for that of a dilute suspension.† For uniform relative motion between a fluid and a solid particle the standard drag curve (drag coefficient C_D vs. Reynolds number N_{Re} based on particle diameter)[36] can be represented by

$$C_D = f_D/\tfrac{1}{2}\pi a^2\bar{\rho}(\Delta U)^2 = 24/N_{Re} \tag{3.5}$$

† See also the theoretical discussion in chapters 5 and 6 of volume 1 of this series.

closely for $N_{Re} \gtrsim 1$, the range of the Stokes law, and approximately for $N_{Re} \gtrsim 700$, where f_D is the drag force, (ΔU) is the relative speed, $N_{Re} = 2a(\Delta U)\bar{\rho}/\bar{\mu}$, $\bar{\mu}$ is the viscosity of the fluid material. For $N_{Re} \lesssim 700$, $C_D \sim 0.44$ was given by Newton.

For particles whose dimensions are smaller than or similar to the mean free path λ of the gas of the suspension, a modification of eqn. (3.5) was suggested for slip flow of molecules with $C_D = 24/N_{Re}C$, where C is a Cunningham correction factor:(37)

$$C = 1+1.6\times 10^{-5}N_{Kn}, \tag{3.6}$$

where $N_{Kn} = \lambda/2a$, the Knudsen number. The approximation of eqn. (3.6) is useful for small N_{Kn}. For sufficiently large N_{Kn}, an accurate approximation is one modifying the drag coefficient for free molecular flow over a sphere, C_{Dfm}, or(38)

$$C_D/C_{Dfm} = 1-[B(S)/N_{Kn}], \tag{3.7}$$

where the speed ratio $S = \Delta U/\sqrt{(2\bar{R}T_\infty)}$, $\bar{R}$ is the gas constant, T_∞ is the static temperature of the fluid away from the sphere, $B(S)$ is a function which ranges from 0.149 to 0.156 for S from 10^{-5} to 0.7, and 0.156 to 0.148 for S from 0.7 to 1; the data are well substantiated by experiments of Millikan.(38) C_{Dfm} for diffuse reflection of molecules and temperature T_w of the surface of the sphere is given by(39)

$$C_{Dfm} = \frac{2}{S^3}\left[\frac{4S^4+4S^2-1}{4S}\,\mathrm{erf}\,S + \frac{e^{-S^2}}{\sqrt{\pi}}(S^2+\tfrac{1}{2})\right] + \frac{2}{3}\frac{\sqrt{\pi}}{S}\sqrt{\left(\frac{T_w}{T_\infty}\right)}. \tag{3.8}$$

A spherical particle in a shear layer of a fluid experiences a lift force f_L. This effect of a velocity gradient was studied by Saffman(40) whose results led to a lift coefficient

$$C_L = f_L/\tfrac{1}{2}\pi a^2\bar{\rho}U^2 = 2K\pi^{-1}[(2a/U)(dU/dy)]^{1/2}(N_{Re})^{-1/2}, \tag{3.9}$$

where the y-direction is normal to the velocity U along the center line, $N_{Re} = 2aU\bar{\rho}/\bar{\mu}$, and $K = 6.46$ is a constant.

When a spherical particle is in a static pressure gradient, the force in the opposite direction of the pressure gradient $\partial P/\partial x$, where P is the pressure, x is the position coordinate, is given by(41)

$$f = -(4\pi/3)(a^3)(\partial P/\partial x). \tag{3.10}$$

As a special case of flow around a sphere in a circular tube, Smythe(42) computed the increase in scalar velocity potential between the ends of a tube with a centrally located sphere; results were expressed in terms of effective increase in tube length.

Elrick and Emrich(43) used smoke particles 0.3 μ as tracers to study flow behavior to within 2 μ from a wall. Fluctuation of 0.15 mm/sec was found in a flow velocity of 50 mm/sec. Their measurements were made in the laminar flow range with a maximum duct Reynolds number of 1100.

Other effects are those due to temperature gradient, non-uniform radiation, and concentration gradient.(1, 44) A recent review has been published in this series by Hidy and Brock.

Free motion of non-spherical particles in a uniform fluid field (such as dropping a rod into a quiescent fluid) involves both translation and oscillation; extensive treatment of motion of such bodies at low Reynolds number is given in the treatise of Happel and Brenner;(45) measurements of forces and moments were made by Marchildon, Clamen, and Gauvin.(λ6) General agreement at low Reynolds number of relative motion was demonstrated.

For a free spherical particle in a locally uniform flow field, its rate of change of momentum is given by, for $\bar{\rho}_p >> \bar{\rho}$,

$$m d\mathbf{U}_p/dt_p = \boldsymbol{f}_D = \tfrac{1}{2}\pi a^2\bar{\rho}C_D|\mathbf{U}-\mathbf{U}_p|(\mathbf{U}-\mathbf{U}_p), \tag{3.11}$$

where m is the mass of the particle, $\mathbf{U}$ and $\mathbf{U}_p$ are the vectorial velocities of the fluid and the particle respectively, and for time coordinate t,

$$d/dt_p = (\partial/\partial t) + \mathbf{U}_p \cdot (\partial/\partial \mathbf{r}), \tag{3.12}$$

where $\mathbf{r}$ is the position vector. Introducing the time constant (inverse relaxation time) of momentum transfer F such that

$$d\mathbf{U}_p/dt_p = F(\mathbf{U}-\mathbf{U}_p), \tag{3.13}$$

we have, with $m = (4\pi/3)a^3\bar{\rho}_p$,

$$F = (3/4)C_D(\bar{\rho}/\bar{\rho}_p)|\mathbf{U}-\mathbf{U}_p|/2a \tag{3.14}$$

and $F = (9/2)\bar{\mu}/\bar{\rho}_p a^2$ for the range where Stokes law applies. It is noted that $\mathbf{U}$ is the mean velocity of the flow field; velocity distribution around the spherical particle is accounted for by F. The general situation of a free particle in a turbulent flow field was formulated by Tchen,[41] who synthesized the results of Basset, Boussinesq, and Oseen to give

$$m(d\mathbf{U}_p/dt_p) = Fm(\mathbf{U}-\mathbf{U}_p) - (4\pi/3)a^3(\partial P/\partial \mathbf{r}) + \tfrac{1}{2}m(\bar{\rho}/\bar{\rho}_p)(d/dt_p)(\mathbf{U}-\mathbf{U}_p) + \boldsymbol{f}_e$$
$$+ 6a^2(\pi\bar{\rho}\bar{\mu})^{1/2}\int_{t_{po}}^{t_p}(t_p-\tau)^{-1/2}(d/d\tau)(\mathbf{U}-\mathbf{U}_p)d\tau, \tag{3.15}$$

where the third term of the right-hand side is the force to accelerate the apparent mass of the particle (significant when the condition $\bar{\rho}_p \gg \bar{\rho}$ is not valid), $\boldsymbol{f}_e$ includes forces due to external field, and the last term is the Basset force due to deviation of flow field from steady state. It is noted that over the spectrum of turbulence, wavelengths greater than $2a$ contribute mainly to the Basset force [last term of eqn. (3.15)], and those smaller than $2a$ contribute mainly to the drag force or time constant F. Most of the earlier studies of particulate suspension attempted solution of eqn. (3.15) followed by an integration procedure. It was identified that when dealing with particles suspended in a turbulent fluid, unless the relation between Eulerian and Lagrangian correlations is known, rigorous solution of eqn. (3.15) cannot be carried out. Soo further pointed out the fact that eqn. (3.15) is, in reality, a transport equation with coordinates $\mathbf{r}$, $\mathbf{U}$, t in the phase space. Therefore, rather than waiting for the knowledge on correlations to develop, problems on dynamics of suspensions can be solved by formulating equations of continuum mechanics. In this way, the coordinates in the phase space are replaced by those in configuration space ($\mathbf{r}$, t) with the introduction of pertinent transport properties.[47]

3.2. Heat transfer by convection from a solid particle

Heat transfer between fluid and a spherical particle by convection is generalized in terms of the Nusselt number, N_{Nu}, and a typical relation is[1]

$$N_{Nu} = 2ah/\bar{\kappa} = 2 + \tfrac{1}{2}N_{Re}N_{Pr} + \ldots, \tag{3.16}$$

where h is the heat transfer coefficient, $\bar{\kappa}$ is the thermal conductivity of the fluid material, $N_{Pr} = c\bar{\mu}/\bar{\kappa}$ is the Prandtl number, c is the specific heat at constant pressure of the fluid. Further, the product $N_{Re}N_{Pr} = N_{Pe}$ is the Peclet number. It is seen that $N_{Nu} = 2$ for a quiescent fluid and is nearly true for relative motion in the Stokes law range.†

† For further discussion, see volume 1 of this series, especially chapters 5 and 6.

Recent contributions on heat transfer from a sphere in a rarefied gas were made by Taylor[48] for low Reynolds number slip flow, and Springer and Tsai[49] gave a method for calculating heat conduction from spheres. Kassoy, Adamson, and Messiter[50] gave, for compressible low Reynolds number flow around a sphere,

$$C_D = 12N_{Re}^{-1}\left[\frac{16C_2(K)}{3K} - \frac{K}{3}\right], \tag{3.17}$$

where $K = \tau(\tau+2)$, $T_w/T_\infty = 1+\tau$, and the functions $C_2(K)$ which vary from 0.042 for $\tau = 0.095$ to 2.28 for $\tau = 1.236$ were tabulated, and

$$N_{Nu} = [(\tau+2)/2(\tau+1)][2+N_{Pr}N_{Re}+0(N_{Re})^2]. \tag{3.18}$$

When dealing with small solid particles whose material has a thermal conductivity $\bar{\kappa}_p$, and when $\bar{\kappa}_p \gg \bar{\kappa}$, uniform temperature over the whole particle can be assumed. The rate of heat exchange by convection is then

$$c_p m(dT_p/dt_p) = 4\pi a^2 h(T-T_p), \tag{3.19}$$

where c_p is the specific heat of the particle, and T and T_p are the static temperatures of the gas and the particle respectively. We introduce time constant for heat exchange G such that

$$dT_p/dt_p = G(T-T_p) \tag{3.20}$$

and

$$\begin{aligned} G &= 3N_{Nu}\bar{\kappa}/2c_p\bar{\rho}_p a^2 \\ &= 3N_{Nu}(\bar{\kappa}/\bar{\kappa}_p)(\bar{\kappa}_p/c_p\bar{\rho}_p)/2a^2, \end{aligned} \tag{3.21}$$

noting that $\bar{\kappa}_p/c_p\,\bar{\rho}_p$ is the thermal diffusivity of the particle material. Unless the fluid contains electrons and ions, F and G are unaffected by the charges carried by the particles.

3.3. Modifications for uncharged liquid droplets

When applied to liquid droplets, the previous relations have to be modified to account for the possibilities of deformation and internal motion within the droplet.

To account for internal motion in a spherical liquid droplet, the Hadamard–Rybcynski relation applicable to $N_{Re} = 2a(\Delta U)\bar{\rho}/\bar{\mu} < 6$[51] is given by

$$C_D = (24/N_{Re})(\tfrac{2}{3}\bar{\mu}+\bar{\mu}_p)/(\bar{\mu}+\bar{\mu}_p), \tag{3.22}$$

where $\bar{\mu}_p$ is the viscosity of the material constituting the liquid droplet. This relation shows that for liquid droplets in air, internal circulation is negligible. For large N_{Re}, Levich[52] postulated that the tangential stress vanishes at the interface and gave

$$C_D = 48/N_{Re}.$$

Chao[53] applied the boundary layer approximation to cases of intermediate N_{Re} and improved the above relation to:

$$C_D = (32/N_{Re})[1+2(\bar{\mu}_p\bar{\mu}^{-1})-0.314(1+4\bar{\mu}_p\bar{\mu}^{-1})N_{Re}^{-1/2}]. \tag{3.23}$$

Measurements by Ingebo[54] on instantaneous drag coefficients for droplets and solid particles of 20 to 120 μ diameter in air yielded

$$C_D = 27N_{Re}^{-0.84} \tag{3.24}$$

for

$$6 < N_{Re} < 400.$$

Due to the interaction of surface forces and inertia forces, a stationary liquid droplet may execute oscillations from the equilibrium shape of a sphere. Relative motion with the gas deforms a droplet into an ellipsoidal shape with a decrease in drag force. At a higher velocity, droplet oscillations may exist, and a large droplet may break up. To account for deformations, the drag coefficient is deformed by a projected area πa_e^2, where a_e is the radius of a sphere of an equivalent volume of the droplet, or

$$C_D = (\text{drag force})/\tfrac{1}{2}\bar{\rho}(\Delta U)^2 \pi a_e^2. \tag{3.25}$$

Relations including the effects of distortion of a droplet, internal circulation, gravity effect, vibration, heat, and mass transfer have been investigated.[1]

When dealing with a water droplet below 1 mm radius in a free fall, Mason[3] suggested that drag coefficient of a rigid sphere is valid.

Winnikow and Chao[55] studied droplet motion in pure liquids and extended the method of Chao[53] to include the effects of internal circulation, deformation, separation of boundary layer, and wake configurations. A modification of eqn. (3.23) was given to include viscous and pressure drags. Droplet deformation was shown to correlate with the product of Weber number $N_{We} = 2a_e(\Delta U)^2(\bar{\rho}_p - \bar{\rho})/\sigma_s$ and Froude number $N_{Fr} = (\Delta U)^2/2a_e g$, where σ_s is the surface tension and g is the gravitational acceleration. Extensive experimental results were obtained to support their theoretical relations.

The influence of interfacial mobility on heat transfer was studied by Conkie and Savic[56] using Pohlhausen's laminar boundary layer velocity profile. They showed that the Stanton number is

$$N_{St} = \frac{C_f}{2}(1 - k_u)^{-1/2}, \tag{3.26}$$

where $N_{St} = h/(\Delta U)\bar{\rho}c$, C_f is the skin fraction factors for an equivalent rigid sphere, and the factor k_u accounts for the interfacial velocity,

$$k_u = -0.0094 N_{Re}(\bar{\mu}/\bar{\mu}_p)^2[1 + (1 + 213 N_{Re}^{-1}\bar{\mu}_p^2\bar{\mu}^{-2})^{1/2}]. \tag{3.27}$$

A thorough survey on the momentum, mass, and heat transfer from deformable particles was given by Winnikow and Chao.[55] A general study of slow, forced scalar transfer from falling droplets was presented by O'Brien.[57]

3.4. Charged liquid droplet

Charge collection (Chapter 2) by a liquid droplet follows similar relations as a solid sphere except that deformation of a spherical droplet may be expected. This is because of the interaction of surface tension and electrostatic repulsion of charges. The ratio of these forces may be denoted as the electrosurface number N_{es}, and

$$N_{es} = 2(q^2/2\varepsilon_0 a)/4\pi a^2 \sigma_s. \tag{3.28}$$

Rayleigh[58] found that a conducting spherical droplet is stable for $N_{es} < 4$. Ailam and Gallily[59] extended this stability criterion to cases of ellipsoids of revolution. Denoting the semi-principal axis along the axis of symmetry as c and the semi-principal axis normal to the axis of symmetry as b, the case where $b > c$ is an oblate ellipsoid and where $b < c$ is a prolate ellipsoid. The radius a_e of the equivalent spherical drop ($b = c = a_e$) is given by $a_e = (bc^2)^{1/3}$. Letting $x = c/b$ and $G^*\pi a_e^2\sigma_s$ be the potential energy of the droplet, we get, for $x < 1$ (oblate ellipsoid),

$$G_1^* = x^{-2/3}\left[2 + \frac{x^2}{(1-x^2)^{1/2}}\ln\frac{1+(1-x^2)^{1/2}}{1-(1-x^2)^{1/2}}\right] + 2N_{es}\frac{x^{1/3}}{(1-x^2)^{1/2}}\tan^{-1}\frac{(1-x^2)^{1/2}}{x}, \tag{3.29}$$

and, for $x > 1$ (prolate ellipsoid),

$$G_2^* = 2x^{-2/3}\left[1+\frac{x^2}{(x^2-1)^{1/2}}\sin^{-1}\frac{(x^2-1)^{1/2}}{x}\right]+2N_{es}\frac{x^{1/3}}{(1-x^2)^{1/2}}\ln\left[\frac{x+(x^2-1)^{1/2}}{x-(x^2-1)^{1/2}}\right]. \quad (3.30)$$

They showed that $d^2G_2/dx^2 = 0$ at $x = 1$ and $N_{es} = 4$ following Rayleigh, and for each N_{es} there exists, in general, a minimum of G^* in each of the range $1 > x > 0$ and $x > 1$. Ailam and Gallily did not draw any conclusion as to the droplet shape of the lowest energy.

However, the physical state of affairs becomes clear if we present their results as shown in Fig. 3, with G^* vs. x with N_{es}, which is given by the particle charge, material, and size,

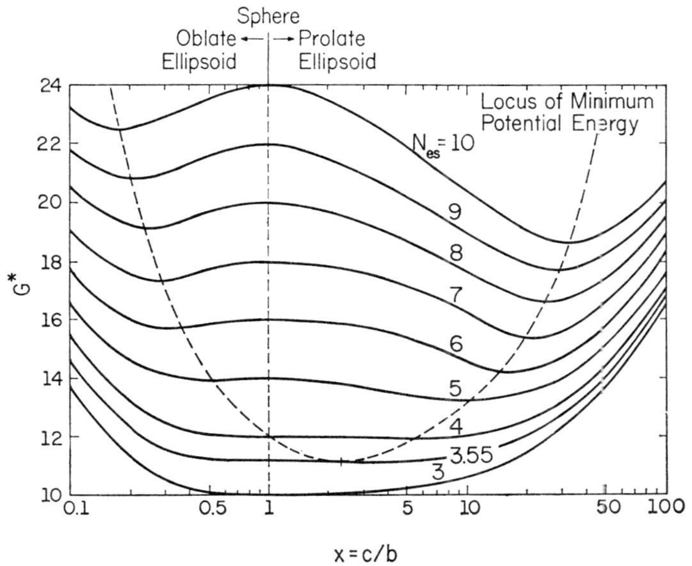

FIG. 3. Dimensionless potential energy of a charged droplet—based on computations by Ailam and Gallily.[59]

as parameter. It is seen that from $x = 1$ to lower values of x, the surface energy increases while the electrostatic energy decreases with a minimum energy at a small x for given N_{es}; from $x = 1$ to larger values of x, the same trend exists except that the x at which minimum energy exists increases with N_{es}, with a value of 2.35 for $N_{es} = 3.55$. The dotted line in Fig. 3 shows the locus of minimum energy. The energy is a minimum at $x = 1$ only for $N_{es} < 3.55$; at $N_{es} = 3.55$ there is one minimum only at $x = 2.35$; at $N_{es} = 4$, $x = 1$ is a state of neutral equilibrium; above $N_{es} = 4$, $x = 1$ are states of unstable equilibrium. This suggests the possibility of non-linear oscillation of a droplet from a prolate to a spherical to an oblate form and back, namely at $N_{es} = 5$ oscillation can occur between $x = 0.15$ and $x = 60$ with $G^* = 15.5$ as a maximum. Such an oscillation, of course, is only the first mode; higher modes would involve representation in terms of zonal harmonics. This same energy diagram also suggests that the minimum on the side $x < 1$ may exist due to fluid interaction, but disturbance would tend to give prolate ellipsoid as the steady droplet shape. Further, when the above oscillation occurs, shattering of a charged droplet tends to occur more often through stretching to prolate ellipsoid or rod shape than by thinning out to an oblate or disk shape. The break-up in the former would be very much similar to a liquid filament and in the latter like a liquid sheet. As a droplet shatters, N_{es} of each droplet would be equal to the original N_{es} divided by the number of similar droplets produced by it.

3.5. Interaction of particles and waves and turbulence

Measurement of snowstorms have been made with light scattering by Mellor.[60] Observation of particulate concentration as an aspect of air pollution control with a ruby Lidar has been experimented by Johnson;[61] calibration appears to be a serious problem.

Additional review of various aspects of attenuation of sound waves by suspended particles and bubbles was given by Blitz.[62]

Further from the above studies and that of Bernard and Binnie,[63] the effect of presence of particles on stream turbulence was studied by Peskin and Wallace using an electrical discharge anemometer.[64] Using alumina shielded platinum electrodes of 0.039 in. diameter, a noise level of 0.002 v RMS was maintained in the glow discharge[15] at 320 v and 3 mA. This discharge probe was installed in a 3 in. by 3 in. flow duct. Comparing the measured turbulence energy spectrum at 85.5 fps flow velocity ($N_{Re} = 66{,}800$) with this discharge probe and with a constant temperature hot wire anemometer, the data obtained with hot wire anemometer was substantiated by the results of Laufer.[65] The microscale Λ was determined to be 0.8 in. Subsequently, this glow discharge anemometer was used to measure the energy spectra when glass beads (0.0041 to 0.0058 in. diameter, $\bar{\rho}_p = 1.5$) were added at the rate of 10.5 lb/min (mass flow ratio: 0.44). The results (power spectrum) are shown in Fig. 4. It was also shown that the intensity of turbulence was decreased by an average of 22% due to the above solid loading. From the experimental results, Peskin and Wallace showed that the Kolmogoroff length scale for the pure gas of η and the suspension η_m has a ratio of 0.77. η is related to the Lagrangian microscale Λ by[65]

$$\Lambda/\eta \sim 15^{1/4}(N_{Re})_\Lambda^{1/2}, \tag{3.31}$$

where

$$(N_{Re})_\Lambda = \langle u^2\rangle^{1/2}\Lambda/\bar{v},$$

$\langle u^2\rangle$ is the intensity, and $\bar{v}$ is the kinematic viscosity of the fluid material.

3.6. Electric drag in an ionized gas

Due to electrostatic interactions, the drag of an ionized gas on a sphere is expected to be different from that of a non-ionized fluid.

Electric drag on a spherical satellite was studied by Jastrow and Pearse,[66] Wyatt,[67] Chopra,[68] and Davis and Harris.[69] Wyatt computed induction drag only by neglecting surface collision but allowing for the shielding effect with the assumption that the body be permeable to electrons and ions. It was shown by Licht[70] that the induction drag due to deflection of particles by electrostatic field alone accounts for about 1% of the total electric drag. Chopra treated the body as a test particle, thus neglecting surface collisions, giving very large values of electric drags. Davis and Harris computed the potential around charged spheres and expressed charge drag in terms of effective radius in number of Debye length. Jastrow and Pearse accounted for the ion-surface collisions; their results are substantiated by the experimental study of Pitts and Knechtel[71] with a mercury plasma.

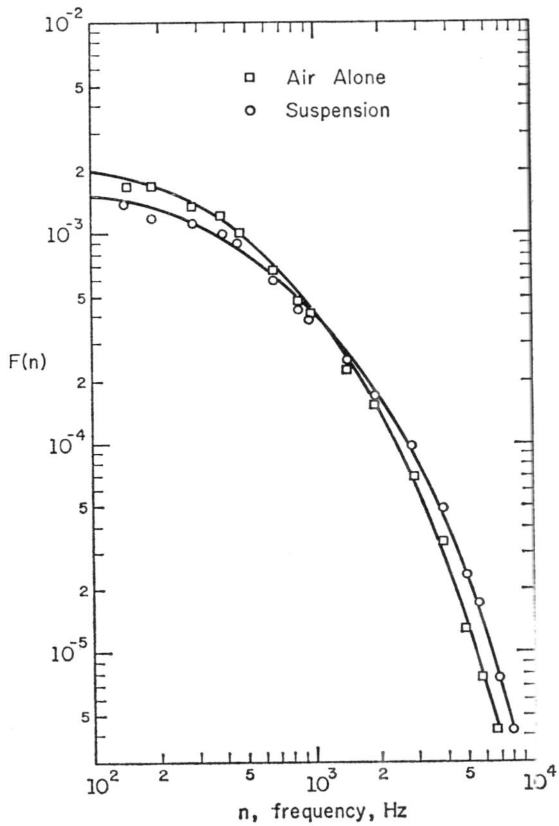

FIG. 4. Effect of particles on the spectrum of turbulence[64] (3 in. by 3 in. duct, 35.5 fps, 0.0041 –0.0058 in. glass particles at 10.5 lb/min in air, mass flow ratio 0.44, $N_{Re} = 16{,}800$).

As in all other studies, Jastrow and Pearse considered high velocity motion of a sphere in a highly ionized gas. They recognized that a sphere of radius a is negatively charged due to greater velocity of electrons. The rate of acceleration of charges leading to Z charges on the sphere in high velocity flow was given by Chopra.[68] A positive ion sheath of thickness δ is formed around the sphere due to repulsion of all but the high energy electrons. It was assumed that δ is uniform, and for surface potential φ_0 due to Z charges at steady state for number density n_i of ions or electrons (no neutrals and with single degree of ionization) δ is given by

$$\delta/a \sim 0.8[-e\varphi_0/(n_i a^2 e^2/2\varepsilon_0)]^{0.47} \tag{3.32}$$

from an approximation to numerical results for

$$0.01 \leq \delta/a \leq 1.0, \tag{3.33}$$

noting that

$$\begin{aligned} -e\varphi_0/(n_i a^2 e^2/2\varepsilon_0) &= -Z/2\pi a^3 n_i \\ &= -2(N_{et})(\lambda_D^2/a^2), \end{aligned} \tag{3.34}$$

where $N_{et} = Ze^2/4\pi\varepsilon_0 akT$ is the electrothermal number and λ_D is the Debye length given by $(\varepsilon_0 kT/n_i e^2)^{1/2}$, with $T = T_e$. Equation (3.33) shows the scaling relation for δ. The electric drag force $\boldsymbol{f}_q$ is given by the following approximation to their numerical results:

$$\frac{\boldsymbol{f}_q}{\boldsymbol{f}_{Di}} = \frac{\boldsymbol{f}-\boldsymbol{f}_D}{{}_{Di}} \sim \left(\frac{e\varphi_0}{E}\right)\left\{1 - \exp\left[-2.4\left(-\frac{e\varphi_0}{E}\right)^{-1}\left(-\frac{2\varepsilon_0 e\varphi_0}{n_i a^2 e^2}\right)^{1/2}\right]\right\} \tag{3.35}$$

for the range

$$0 \leq (-e\varphi_0/E) \leq 1, \tag{3.36}$$

where $\boldsymbol{f}$ is the total drag and $\boldsymbol{f}_{Di}$ is the drag force of an uncharged sphere due to n_i, E is the ion kinetic energy relative to the sphere; $E = (1/2)m_i U^2$. Equation (3.35) was shown by Pitts and Knechtel[71] to be accurate within 10% of their experimental results. Additional scaling relation is seen in that

$$-\frac{e\varphi_0}{E} = \left(-\frac{Ze^2}{4\pi\varepsilon_0 akT}\right)\left(\frac{kT}{(1/2)m_i U^2}\right) = (2/\gamma)(-N_{et})(N_M)^{-2}, \tag{3.37}$$

where γ is the ratio of specific heats and N_M is the Mach number of the sphere based on the speed of sound $\sqrt{(\gamma\bar{R}T)}$, where $\bar{R}$ is based on atom or ion in the gas. Equation (3.35) now becomes

$$\frac{\boldsymbol{f}_q}{\boldsymbol{f}_{Di}} \sim \frac{2}{\gamma}\frac{(-N_{et})}{(N_M)^2}\left\{1 - \exp\left[-1.69\gamma\frac{(N_M)^2}{(-N_{et})^{1/2}}\frac{\lambda_D}{a}\right]\right\} \tag{3.38}$$

for cases where the bracketed term is small or where $(-e\varphi_0/E)$ is large, eqn. (3.38) reduces to the asymptotic value

$$\begin{aligned} {}_q/\boldsymbol{f}_{Di} &\sim 2.4(-2N_{et})^{1/2}(\lambda_D/a) \\ &\sim 2.4(-Z/2\pi n_i a^3)^{1/2} \end{aligned} \tag{3.39}$$

for low relative velocity. It is noted that eqn. (3.35) applies to cases where conditions given by eqn. (3.33) or

$$0.01(-N_{et})^{-1/2} < (\lambda_D/a) < (-N_{et})^{-1/2},$$

that is

$$10^{-4} 2\pi n_i a^3 < Z < 2\pi n_i a^3,$$

and eqn. (3.36), or

$$0 < (-N_{et}) < N_M^2$$

but not arbitrary ion concentration and charge Z of the sphere. Therefore the approximation given by eqn. (3.35) is applicable to motion at high Mach number and relatively large a. In the limit of large N_M,

$$\boldsymbol{f}_q/\boldsymbol{f}_{Di} \sim (2/\gamma)(-N_{et})(N_M)^{-2}. \tag{3.40}$$

The formulation of Jastrow and Pearse is, in effect, an "electric boundary layer" approximation in a highly ionized gas. It does not apply to the case of highly charged spheres in slow motion in a partially ionized gas.

For conditions such as exist in the case of charged fog particles and charged dust in an electrostatic precipitator, low relative velocity and low extent of ionization of the gas must be accounted for. Since $\boldsymbol{f}_{Di}$ accounts for ions only, eqns. (3.5) and (3.6) give $\boldsymbol{f}_{Di} = 6\pi\mu_i aU/C$, where μ_i is due to ions only and C is the Cunningham correction factor. From eqn. (3.39) it is readily seen that the additional drag coefficient due to electric charges is

$$\Delta C_{Dq} = \frac{\boldsymbol{f}_q}{(1/2)\rho U^2 \pi a^2} \sim \frac{14.4}{C}\left(\frac{\lambda}{a}\right)\left(\frac{n_i}{n}\right)^{1/2}\frac{\langle v_i \rangle}{U}\left(-\frac{Z}{2\pi n a^3}\right)^{1/2} \tag{3.41}$$

for $m_i = m$ of gas molecules, where n is the total number density of the gas of density ρ,

$\langle v_i \rangle = \sqrt{(8kT/\pi m_i)}$, λ is the mean free path. The ΔC_{Dq} is to be added to the hydrodynamic drag coefficient for the total drag coefficient. The magnitude of ΔC_{Dq} is seen in these cases: (1) 10 μ fog particle falling at 0.01 m/sec = U, in an atmosphere of 10^{-6} fraction of ionization, eqn. (2.18) gives $N_{et} = -6.37$, $Z = -574$, $\lambda/a = 10^{-2}$, $\Delta C_{Dq} \sim 0.007$; (2) 10 μ particle charged to $Z = -31{,}000$ in an electrostatic precipitator,(14) 10^{-4} fraction ionization, $\lambda/a = 10^{-2}$, $U = 0.1$ m/sec, $\Delta C_{Dq} \sim 5$.

The effect of a magnetic field on a charged sphere in an ionized gas was studied by Hoult,(72) who found that the magnetic field has a stabilizing influence on the motion of a body, although the drag force was not computed.

CHAPTER 4

PARTICLE CLOUDS

Momentum and energy transfer between clouds of particles of different species or sizes has been extensively treated.[(1)] Moreover, interest in atmospheric science concerns mainly dilute suspensions. With respect to interactions among clouds of particles, we shall only discuss recent results in charge transfer, which is also the basis of electrostatic probes for a suspension. This phenomenon influences atmospheric sampling of aerosols and ice crystals.

Another recent result which is believed to have an important bearing on the boundary conditions for a suspension, is the concept of interaction length[(73, 77)] and the state of motion of particle clouds.

A survey of recent studies of condensation, evaporation, droplet coalescence, and break-up will also be included in this chapter.

4.1. Transfer of electric charges

We consider the case of sphere 2 impacted by a cloud of spheres 1 (see section 1.1) From eqn. (1.7) the change of potential of sphere 2 is given by[(5)]

$$C_2 \frac{dV_2}{dt} = \eta_{21} n_1 \Delta u_{21} \int_0^{\pi/2} 2\pi Q_{21} (a_1 + a_2)^2 \sin\theta \, d\theta, \tag{4.1}$$

where η_{21} is the fraction impacted, defined as the ratio of the cross-sectional area of stream of particles 1 to the projected area of particle 2 normal to the direction of flow of particles 1, and n_1 is the number density of particles 1. η_{21} due to fluid and electrostatic forces has been extensively studied.[(1)] Since $C = dQ/dV$, the total electric current due to successive charge transfer is:

$$i_2 = \frac{5}{4} 2\pi (a_2 + a_1)^2 \eta_{21} n_1 \Delta u_{21} (V_2 - V_1) h_{21} (A_{21} \Delta t)_0 \tag{4.2}$$

with mean value of $(V_2 - V_1)$ over Δt.

By analogy to momentum and heat transfer processes,[(1)] the charge transfer process involves an inverse relaxation time H_{21}, with

$$\frac{dV_2}{dt} = -H_{21} (V_2 - V_1). \tag{4.3}$$

Comparison to eqn. (4.1) with substitution of eqns. (1.4) and (1.10) gives

$$H_{21} = K_{21} \frac{a_2^2}{\varepsilon_2} \frac{\rho_1}{\bar{\rho}_1} \frac{h_{21}}{(a_2 a_1)^{1/2}}, \tag{4.4}$$

where K_{21} is a dimensionless constant given by

$$K_{21} = \frac{15}{32}(0.97)\eta_{21}\left[\left(\frac{m_2}{m_1}\right)^{1/2}+\left(\frac{m_1}{m_2}\right)^{1/2}\right]^{-4/5}\left[\left(\frac{a_2}{a_1}\right)^{1/2}+\left(\frac{a_1}{a_2}\right)^{1/2}\right]^{7/5}\times$$
$$\times\left[\left(\frac{k_2}{k_1}\right)^{1/2}+\left(\frac{k_1}{k_2}\right)^{1/2}\right]^{4/5}[(N_{Im})_{21}]^{4/5}.$$

Further, for charge transfer between two clouds of particles of species 1 and 2, we have the constitutive relation

$$C_2 n_2 H_{21} = C_1 n_1 H_{12} \tag{4.5}$$

based conservation of charges. It is seen that the asymptotic relation from eqn. (1.9) is

$$\varphi + \frac{q}{4\pi\varepsilon a} = \text{constant}; \tag{4.6}$$

for each species of particles in a cloud and for a cloud of particles of similar materials,

$$q/a = \text{constant} \tag{4.7}$$

by mutual collision.

The relation given by eqn. (4.2) gives rise to an electrostatic probe theory for a suspension. When an impact probe is inserted into a flow stream of particles, charge transfer occurs by impact. For probe dimensions smaller than the mean free path of particles,[1] single scattering consideration would be adequate. The potential of each particle is given by

$$V_1 = \varphi_1 + \frac{q_1}{4\pi\varepsilon_1 a} \tag{4.8}$$

for adjustable probe voltage V_2.

For a stream of particles (species 1) impacting on a stationary probe in the form of a metal sphere, the current reading from its lead wire is given by eqn. (4.2) for $a_2 \gg a_1$:

$$\begin{aligned} i_2 &= \eta_{21} A\rho_1 u_1 Q_{21}/m \\ &= (129.6)\eta_{21} h_{21} a_2^2 \frac{\rho_1}{\bar{\rho}_1}\left[(\bar{\rho}_1 u_1^2)(k_1+k_2)\left(\frac{1+r^*}{2}\right)\right]^{4/5}(V_2 - V_1), \end{aligned} \tag{4.9}$$

where A is the projected area of the probe and Q_{21} is the charge transferred per impact.

Experiments were performed on a recirculating flow loop with a 5 in. diameter brass pipe. Pulverized coal (Ohio) suspension in air was tested at various air velocities. The coal dust has a normal size distribution with mean at 16 μ.

Figure 5 shows the relation between the ball-probe current and the mass flow for $\frac{1}{2}$ in. steel balls in coal dust suspensions at two flow velocities.[74] Figure 5 shows the effect of probe voltage and impact velocity. V_f is the floating probe potential. At mass flows below 0.2 gm/cm² sec these calibration curves do not follow simple shapes because of the variation of charge to mass ratio at low mass flow.

4.2. Impact on a wall by a turbulent suspension

In the case of a fully developed turbulent suspension flowing in a circular duct, velocities of particles (species 1) hitting the wall are those of its random motion with intensity $\langle v^2\rangle^{1/2}$. We shall treat their directions as random and equally probable. The number of collisions per unit area and time is thus $(1/2)n_1\langle v_1\rangle^{1/2}\cos\theta\sin\theta\, d\theta$, where θ is the angle measured from a normal to the wall (species 2). Analogous to the relations for

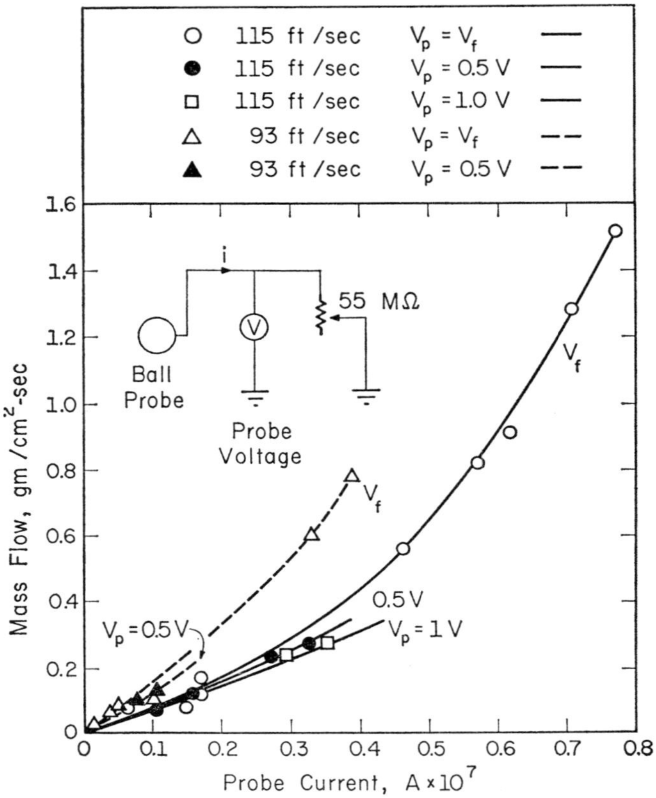

FIG. 5. Electrostatic ball probe current. Measurements made by L. Cheng.

momentum and heat transfer,[1] integrations based on eqns. (1.4) and (1.10) give the current density to the duct wall:

$$\begin{aligned} J_w &= \left(\frac{5}{18}\right) n_1 \langle v_1^2 \rangle^{1/2} h_{21}(V_2 - V_1)(A_{21}\,\Delta t)_0 \\ &= \left(\frac{5}{18}\right)(0.97\pi) n_1 a_1^3 h_{21} \left[\left(\frac{k_2}{k_1}\right)^{1/2} + \left(\frac{k_1}{k_2}\right)^{1/2}\right]^{4/5} [(N_{Im})_{21}]^{4/5} (V_2 - V_1) \\ &= -J_1, \end{aligned} \tag{4.10}$$

where J_1 is the current density to the particle cloud.

For $a_2 \gg a_1$, $m_2 \gg m_1$, (N_{Im}) is given by

$$(N_{Im})_{21} = 5\pi^2 \langle v_1^2 \rangle \bar{\rho}_1 (k_2 k_1)^{1/2} (1 + r^*)/2.$$

J_w can be expressed in terms of particle cloud density ρ_1 with $n_1(4\pi/3)a_1^3 = \rho_1/\bar{\rho}_1$. The inverse relaxation time for charge transfer is

$$H_{21} = K_{21}(h_{21} a_2/\varepsilon_2)(\rho_1/\bar{\rho}_1), \tag{4.11}$$

where

$$K_{21} = \frac{15}{32}(0.97)[(k_2/k_1)^{1/2} + (k_1/k_2)^{1/2}]^{4/5}[(N_{Im})_{21}]^{4/5}.$$

The charge transferred per particle is given by

$$Q_{21} = J_1(6\pi)^{1/2}/n_1\langle v_1^2\rangle^{1/2}, \tag{4.12}$$

which gives, for $h_{21} \sim \sigma_1/a_1$ which applies to insulator particles in a metal duct,

$$Q_{21} \sim 2.2\pi^{13/5}a_1^2\sigma_1(k_1+k_2)^{4/5}\langle v_1^2\rangle^{3/10}(\bar{\rho}_1)^{4/5}(V_2-V_1)\left(\frac{1+r^*}{2}\right)^{4/5}, \tag{4.13}$$

suggesting that the charge gained after a finite number of collisions is nearly proportional to the square of the particle radius, which was experimentally verified before.[75] Due to the small number of collisions, the asymptotic value such as given by eqns. (4.7) and (4.8) was seldomly reached even over a long duct.[76] Recycling system is complicated by the presence of the blower and bends in the experimental system.[5] Further, for constant h_{21}, the bulk charge to mass ratio is independent of particle size a. This has been demonstrated by experimental results.

4.3. Interaction length, slip, and accommodation

In a dilute particulate suspension, the effect of a particle–particle interaction is negligible. However, there is an interaction length L_p of particles with fluid due to Brownian motion, wall interaction, wake effect or fluid turbulence, or their combinations,

$$L_p \sim \langle\Delta u^2\rangle^{1/2}/F, \tag{4.14}$$

where $\langle\Delta u^2\rangle^{1/2}$ is the RMS of relative velocity between the particles and the fluid. L_p may become large when compared to the characteristic physical dimensions of a system. Take the example of flow over a flat plate[73,77] with the flat plate along the x-direction, $x = 0$ at the leading edge, velocities u, u_p of fluid and particles, and free stream velocity U, we have the boundary condition of particle velocity at the wall (y normal to x):

$$u_{pw} = U(1-\xi)+L_p(\partial u_p/\partial y)_{y=0}, \tag{4.15}$$

where $\xi = Fx/U$, and $\xi = 1$ for $Fx/U > 1$. Analogous to the case of a rarefied gas we have an interaction Knudsen number

$$N_{Kp} = L_p/x. \tag{4.16}$$

For similar reasons, the interaction length for convective heat transfer between the particle and the fluid is

$$L_T = \langle\Delta u^2\rangle^{1/2}/G = L_p(F/G). \tag{4.17}$$

For the above example with temperatures T_w at the wall and T_∞ in the free stream, we have the boundary condition for particle temperatures given by

$$T_{pw}-T_w = (T_\infty-T_w)(1-\xi_T)+L_T(\partial T_p/\partial y)_{y=0} \tag{4.18}$$

with $\xi_T = Gx/U = \xi F/G$ and for $\xi_T = 1$ at $Gx/U > 1$; T_{pw} is the temperature of the particles at the wall. It is further noted that, since the temperature of the particle phase is that of the body temperature of the particles, heat conduction by surface contact must be accounted for.[77] For elastic collision of particles with a wall, it is readily shown that an accommodation coefficient α is given by

$$-\ln(1-\alpha) \sim (15/16)2^{-2/5}(2.94)[(\bar{\kappa}_p\bar{\kappa}_w)^{1/2}/c_p\bar{\rho}_p a(\Delta u_w)](N'_{Im})^{4/5}, \tag{4.19}$$

where the impact number N'_{Im} is given by

$$N'_{Im} = \pi^2(5/2)(\Delta u_w)^2(1+r^*)\bar{\rho}_p k_p[1+(k_p/k_w)], \tag{4.20}$$

where r^* is the ratio of the reflected speed to the incoming speed Δu_w, $k = (1-\nu^2)/\pi E$, ν is the Poisson ratio, E is the modulus of elasticity. The temperatures of an incoming particle T_{pi} and a reflected particle T_{pr} are related by

$$(T_{pr}-T_{pi})/(T_w-T_{pi}) = \alpha. \tag{4.21}$$

Since $T_{pw} \sim (T_{pi}+T_{pr})/2$, the heat flux J_{qp} due to collision of particles at the wall is given by

$$J_{qp}/c_p\dot{m}_{pw} = (T_{pr}-T_{pi}) = [2\alpha/(2-\alpha)](T_w-T_{pw}), \tag{4.22}$$

where $\dot{m}_{pw}$ is the mass flux of impact at the wall. Equations (4.18) and (4.22) account for both the effect of conduction by surface contact at the wall and the effect of convection heat transfer with the fluid.

4.4. State of motion of particles

For a dilute suspension the fluid properties—viscosity $\bar{\mu}$, thermal conductivity $\bar{\kappa}$, diffusivity D, and state of motion of the fluid—are unaffected by the presence of the particles.

Since the effect of particle–particle collisions is negligible in a dilute suspension, the motion of particles is not correlated among themselves. The effect of particle–fluid interaction predominates in the state of motion of particles. By analogy to the rarefied state of a gas, the interaction length L_p given by eqn. (4.14) in relation to the physical dimension of a flow system gives a measure of the relative influence of interaction of particles with a solid boundary and their interaction with the fluid. L_p is independent of ρ_p. Take the case of pipe (or radius R) flow of a suspension when the fluid is turbulent for instance, R/L_p gives the ratio of number of particle–vortex interactions to particle–wall collisions. For $R = 2.5$ in., for 100 ft/sec flow velocity in the core and $L_p = 0.6$ in., this ratio is nearly 4 as compared to 7×10^5 for molecules of air at STP. Hence, even though the fluid is turbulent, the motion of the particles is in a regime corresponding to a rarefied state of a gas with their random motion sustained by fluid turbulence. The existence of L_p explains the experimental result that the particles are in slip motion at the pipe wall.

When the fluid is in laminar motion, L_p is the free path over which a particle changes its direction because of Brownian motion, collision with a wall, or the wakes of other particles. The latter arises in laminar motion when the particle size is large compared to the thickness of the shear layer. These conditions are illustrated in Chapter 8.

In the regime of laminar motion of the fluid, in the limit, D_p, the particle diffusivity, is the Brownian diffusivity

$$D_p = 2kT/6\pi\bar{\mu}a, \tag{4.23}$$

where k is the Boltzmann constant and T is the absolute temperature. In a suspension in a turbulent fluid, D_p is the diffusivity of the particles in the mixture and is related to the eddy diffusivity of the fluid and parameters of fluid–particle interaction.[(77)] It is noted that random motion of particles exists in a suspension when the fluid is in laminar motion in the same way as random molecular motion exists in the fluid itself. Moreover, molecules in a glass of water at rest have random motion, and the water is certainly not in a state of turbulence.

In the shear motion of a dilute suspension, the shear stress in the fluid is contributed by its viscosity $\bar{\mu}$, while the particles diffuse through the shear layer, or the "shear stress" τ_p of the particles is given by, for simple shear motion (gradient of momentum transfer)

$$\tau_p = \mu_p \partial u_p/\partial y \sim \rho_p D_p \partial u_p/\partial y, \tag{4.24}$$

where $\partial u_p/\partial y$ is the shear rate with velocity u_p and normal distance y. Turbulent motion of the particle phase exists only when sufficient mutual collision of particles causes their correlated motion.

4.5. Condensation, evaporation, droplet coalescence, and break-up

The multiphase formulation[1] accounts for, as phase change, any change from species p to species q particles by condensation,[1] evaporation, coalescence, or breaking-up.

Condensation of water vapor in the atmosphere may begin on small ions but principally with aerosol (0.1–10 μ) as nuclei.[78] The seeding of super-cooled clouds may also be done with ice nuclei to promote freezing of super-cooled water droplets. Nucleation of ice and rain drops may be produced by organic compounds[79] or by zinc dust.[80] Kinetics of growth and evaporation of droplets and ice crystals were treated by Philip.[81] Growth of ice crystals from vapor was studied by Hallett.[82] Extensive discussion of various steps of condensation in cloud formation was given in treatises by Mason[3] and Fletcher,[78] and in recent studies.[81,82] However, knowledge on basic nucleation mechanism is still incomplete although once a droplet is formed subsequent condensation process is quite well known.[1,3] Other mechanisms are charge transfer associated with the evaporation of ice,[83] collision velocity, air pressure, and charge transfer.

For small droplets electric forces have significant effects on the formation of precipitation as formulated by Krasnogorskaja.[10] Luminosity produced by coalescence of oppositely charged falling water droplets has been observed.[84]

The dynamics of coalescing or shattering of particles are not accounted for by the relaxation mechanism of particle–particle interaction[1] but are accounted for in terms of accommodation of generated species. Physically, it is seen that when a particle 1 and a particle 2 collide and combine, the momentum of the particle 3 generated is given by conservation of momentum as (m for mass and U for velocity)

$$m_3U_3 = m_1U_1+m_2U_2 = (m_1+m_2)U_3. \tag{4.25}$$

The energy dissipated is given by

$$(1/2)m_1U_1^2+(1/2)m_2U_2^2-(1/2)m_3U_3^2 = (1/2)m_1m_2(m_1+m_2)^{-1}(U_1-U_2)^2, \tag{4.26}$$

which might produce a rise in temperature or shattering of the drop. Since the details of such interactions can become extremely complicated, approximation by continuum mechanics is again desirable.

A simple case of growth of a large droplet (accretion) in a fog was treated by Mason.[3] A water drop larger than neighboring droplets acquires also a larger terminal velocity. This results in its collision with the smaller neighbors lying in its path. If a fraction η_{c0} of the droplets collided (with fraction impacted η_{12}) with the larger droplet coalesce with it, the larger droplet (of radius a_1) grows at a rate given by

$$\frac{dm_1}{dt} = 4\pi a_1^2\bar{\rho}_p da_1/dt = \eta_{12}\eta_{co}\pi a_1^2\rho_{p2}(U_{p1}-U_{p2})$$

or

$$\frac{da_1}{dt} = \eta_{12}\eta_{co}(\rho_{p2}/\bar{\rho}_p)(U_{p1}-U_{p2}); \tag{4.27}$$

for $U_{p1} \gg U_{p2}$ and $U_{p1} = -dy/dt$ we get

$$\frac{da_1}{dy} = -\eta_{12}\eta_{co}(\rho_{p2}/\bar{\rho}_p) \tag{4.28}$$

for the growth of droplet size with altitude y of fall.

In this case, the rate of change of momentum of the droplet (1) is influenced by coalescence with droplets (2) according to

$$d(m_1 U_{p1}) = U_{p2}\, dm_1, \tag{4.29}$$

assuming mass m_2 remains constant. Substitution of eqn. (4.27) gives the momentum equation of falling droplet (1), whose mass is changing, as (from eqn. 5.5)

$$\frac{dU_{p1}}{dt} = -g - F_1 U_{p1} + F_{p12}(U_{p2} - U_{p1}) \tag{4.30}$$

in still atmosphere; and F_{p12}, the relaxation time for particle–particle interaction, is given by[1]

$$F_{p12} = (3/4)\eta_{12}\eta_{co}(\rho_{p2}/\bar{\rho}_p)|U_{p2} - U_{p1}|/a_1. \tag{4.31}$$

For a large drop, $C_D = 0.44$ at $N_{Re} \gg 1$, eqn. (3.14) gives

$$F_1 = (3/8)C_D(\bar{\rho}/\bar{\rho}_p)|U_{p1}|/a_1. \tag{4.32}$$

For the case of steady motion with $|U_{p1}| \gg |U_{p2}|$ and constant η_{12}, eqn. (4.30) with (4.28) gives (subscript 0 for initial conditions)

$$(U_{p1})^2 = (ga_1/F')2[2+(\alpha'/F')]^{-1}[1-(a_{10}/a_1)^{2(F'/\alpha')+1}] + (U_{p01})^2(a_{10}/a_1)^{2(F/\alpha')} \tag{4.33}$$

with $F' = (3/8)C_D(\bar{\rho}/\bar{\rho}_p) + (3/4)\eta_{12}\eta_{c0}(\rho_{p2}/\bar{\rho}_p)$ and $\alpha' = \eta_{12}\eta_{c0}\rho_{p2}/\bar{\rho}_p$. The coordinate system requires that $U_{p1} < 0$. For U_{p1} at the terminal value to begin with, $(U_{p01})^2 = ga_{10}/F'$, we have

$$(U_{p1})^2 = (ga_1/F')2[2+(\alpha'/F')]^{-1}[1-(a_{10}/a_1)^{2(F'/\alpha')+1}] + (a_{10}/a_1)^{2(F'/\alpha')+1} \tag{4.34}$$

and

$$(a_1/a_{10}) = 1 + (\eta_{12}\eta_{co}y_o/a_{10})(\rho_{p2}/\bar{\rho}_p)[1-(y/y_o)]. \tag{4.35}$$

It is noted that at $\alpha' = 0$, $a_1 = a_{10}$ always, and we have the case of free fall of drop (1) in a gas. The case $(F'/\alpha') \to (3/4)$ represents the case of dense fog, where the velocity of fall of (1) is influenced by coalescence alone. It is seen that all finite values of F' and α' reduce the terminal velocity below that in a gas as expected. For an example of $C_D = 0.44$, $a_1 = 1$ mm, $y_0 = 30$ m, $\rho_{p2} = 10^{-3}$ kg/m^3, $\bar{\rho}_p = 10^3$ kg/m^3, $\bar{\rho} = 1$ kg/m^3, $\eta_{12}\eta_{c0} = 1/4$, $F' = 0.165 \times 10^{-3} + 0.2 \times 10^{-6}$, $\alpha' = 0.24 \times 10^{-6}$, at $y = 0$, $a_1 = 1.0075$ mm, $U_{p1} = 7.74$ m/sec $\sim U_{p01}$ ($N_{Re} = 885$) (note that Gunn and Kinzer[3] gave 6.49 m/sec). Size and size distribution of raindrops in a thunderstorm vary over wide ranges.[85,86]

Shattering of a liquid jet and droplets by a stream of air was extensively studied by Lane[87] and Clark.[88] However, it appears that not much is known about shattering of droplets by mutual collision. The electrostatic effect on droplet break-up is demonstrated in Chapter 3. Drops may even freeze due to drop break-up.[90] Extensive studies have been made on sonic agglomerations and dispersion of particles and droplets.[89]

For the case involving coagulation only, Twomey[91] computed the rate of rain formation by integrating the statistical relation describing the droplet size distribution:

$$(d/dt)n(\mathbf{v}_1) = -n(\mathbf{v}_1)\int_0^\infty K(\mathbf{v}_1, \mathbf{v})n(\mathbf{v})\,d\mathbf{v} + \tfrac{1}{2}\int_0^{\mathbf{v}_1} K(\mathbf{v}, \mathbf{v}_1-\mathbf{v})n(\mathbf{v})n(\mathbf{v}_1-\mathbf{v})\,d\mathbf{v} \tag{4.36}$$

for the rate of change of number density of droplets of volume $\mathbf{v}_1$, where $K(\mathbf{v}_1, \mathbf{v})$ is the coagulation coefficient for number density of droplets of volume $\mathbf{v}_1$ coagulating with that of droplets of volume $\mathbf{v}$. The first term on the right-hand side gives the number of $\mathbf{v}_1$ lost by increase in size while the second term gives the number of drops of $\boldsymbol{v}_1$ generated by coagulation. It was shown that clouds with 50 droplets/cm^3 produce rain drops of 100–400 μ in 20–30 min.

4.6. Mobility and conductivity

For particles of charge to mass ratio q/m, its mobility is defined by its relative velocity to the uncharged principal fluid

$$\Delta U = [(q/m)/F]E \equiv K_p E. \tag{4.37}$$

For given mobilities of components of a mixture, its conductivity is given by

$$\sigma = \sum_Z q_z n_z K_{pz}. \tag{4.38}$$

Fair weather conductivity due to particle pollution was determined by Gunn[(92)] who showed that this conductivity has increased by 5% between 1929 and 1962.

CHAPTER 5

HYDRODYNAMIC EQUATIONS OF A COMPONENT IN A MIXTURE

The continuum mechanics of a single-phase fluid amounts to a successful simplification of the general transport equation of kinetic theory of a flow system by replacing the coordinates in the phase space (position and momentum coordinates) with configuration space (position coordinates) through the introduction of pertinent phenomenological relations and transport properties. This rationale is applicable to the dynamics of a suspension because we are not always interested in the details of the path of an individual aerosol; rather the group behavior of a cloud of aerosols is what we are usually after.

It has been demonstrated[1,93] that when dealing with a particulate system or a gaseous mixture of molecules of disparate masses, continuum mechanics can be extended to treat the motion of an individual component or species interacting with other components of the mixture. Visualizing a mixture as a number of pure components occupying the same volume in space, the density of a mixture ρ_m is given by

$$\rho_m = \sum_{(q)} \rho^{(q)}, \tag{5.1}$$

where $\rho^{(q)}$ is the density of component (q) of the mixture occupying the volume of the mixture of ν components $(q = 1,\ldots,\nu)$. The ith component of velocity of the mixture U_{mi} is given by

$$\rho_m U_{mi} = \sum_{(q)} \rho^{(q)} U_i^{(q)}. \tag{5.2}$$

5.1. Continuity equation

Taking the general case, including chemical reaction, the continuity equation of component (q) of the mixture can be expressed as

$$\partial\rho^{(q)}/\partial t + (\partial/\partial x_j)(\rho^{(q)} U_j^{(q)}) = \Gamma^{(q)}, \tag{5.3}$$

where t is the time, x_j is the jth component of the space coordinate, $\Gamma^{(q)}$ is the rate of generation of (q) per unit volume, and $\Sigma_{(q)}\, \Gamma^{(q)} = 0$ over the whole mixture.

For a reactive system,

$$\Gamma^{(q)} = m^{(q)} \sum_j \frac{dn_j^{(q)}}{dt} = m^{(q)} \sum_j k_j \Pi_{(p)} (n_j^{(p)})^{\nu_j(p)}; \tag{5.4}$$

by summing over reactions j, reactions rates $k_j > 0$ for backward reactions and $k_j < 0$ for forward reactions, m denotes the mass, n is the number of moles, and ν_j denotes the stoichiometric coefficient.

5.2. Momentum equation

Illustrating with a Newtonian fluid, it is readily shown that the momentum equation of component (q) as a pure continuum is given by

$$\rho^{(q)}dU_i^{(q)}/dt^{(q)} = (\partial/\partial x)_j[-P^{(q)}\delta_{ji}+\mu_m^{(q)}\Delta_{ji}^{(q)}+\mu_{m2}^{(q)}\Theta^{(q)}\delta_{ji}]+ +\rho^{(q)}F_i^{(q)}-(U_i^{(q)}-U_{mi})\Gamma^{(q)}+\rho^{(q)}\sum_{(p)}F^{(qp)}(U_i^{(p)}-U_i^{(q)}), \quad (5.5)$$

where the total derivative

$$d/dt^{(q)} = (\partial/\partial t)+U_j^{(q)}(\partial/\partial x_j).$$

$P^{(q)}$ is the partial pressure of component (q), $F_i^{(q)}$ is the ith component of force acting as a unit of mass of (q), $F^{(qp)}$ is the time constant (inverse of relaxation time) of momentum transfer between components (q) and (p). The deformation tensor and dilatation are

$$\Delta_{ji}^{(q)} = (\partial U_i^{(q)}/\partial x_j)+(\partial U_j^{(q)}/\partial x_i), \quad (5.6)$$

$$\Theta^{(q)} = \tfrac{1}{2}\Delta_{kk}^{(q)} = (\partial U_k^{(q)}/\partial x_k). \quad (5.7)$$

Particular attention should be given to the viscosity of component (q) in the mixture $\mu_m^{(q)}$ which is related to the viscosities of interactions of components $\mu^{(qq)}$, $\mu^{(qp)}$, $\mu^{(pp)}$, etc., which constitute μ_m of the mixture. An approximation was made to determine $\mu_m^{(q)}$ from μ_m and $\mu_{m2}^{(q)}$ from μ_{m2}, where $\mu_{m2} = \zeta_m-(2/3)\mu_m$, ζ_m being the bulk viscosity of the mixture. As an example, $\mu_m^{(q)}$ is given by[94]

$$\mu_m^{(q)}\Delta_{ji}^{(q)} = \mu_m(\rho^{(q)}/\rho_m)\Delta_{ji}^{(q)}+(\mu_m/\rho_m)[(U_i^{(q)}-U_{mi})(\partial\rho^{(q)}/\partial x_j)+(U_j^{(q)}-U_{mj})(\partial\rho^{(q)}/\partial x_i)]+ +\rho^{(q)}(U_i^{(q)}-U_{mi})(U_j^{(q)}-U_{mj}). \quad (5.8)$$

Further, the interaction of components is such that

$$\sum_{(q)}\sum_{(p)}\rho^{(q)}F^{(qp)}(U_i^{(p)}-U_i^{(q)}) = 0. \quad (5.9)$$

5.3. Energy equation

The energy equation of component (q) can be expressed as

$$\rho^{(q)}dE^{(q)}/dt^{(q)} = (\partial/\partial x_j)U_j^{(q)}[-P^{(q)}\delta_{ji}+\mu_m^{(q)}\Delta_{ji}^{(q)}+\mu_{m2}^{(q)}\Theta^{(q)}\delta_{ji}]-\Gamma^{(q)}E^{(q)}+ +(\partial/\partial x_j)(\kappa_m^{(q)}\partial T^{(q)}/\partial x_j)+G^{(q)}+c^{(q)}\rho^{(q)}\sum_{(q)}G^{(qp)}(T^{(p)}-T^{(q)}), \quad (5.10)$$

where the energy $E^{(q)} = u^{(q)}+\frac{1}{2}U^{(q)2}$, $u^{(q)}$ is the internal energy of component (q)† the relation to the energy of the mixture is given by

$$\rho_m E_m = \sum_{(q)}\rho^{(q)}E^{(q)} \quad (5.11)$$

and specific heats at constant volume are related according to

$$\rho_m c_{vm} = \sum\rho^{(q)}c_v^{(q)}. \quad (5.12)$$

$T^{(q)}$ its static temperature, $G^{(q)}$ is the rate of energy generation per unit volume which may include heat of mixing, heat of reaction or phase change; $\kappa_m^{(q)}$ is the thermal conductivity of component (q) in the mixture which is related to κ_m of the mixture. $G^{(qp)}$ is the time constant of heat transfer between components (p) and (q), and over the whole mixture

$$\sum_{(q)}\sum_{(p)}c^{(q)}\rho^{(q)}G^{(qp)}(T^{(p)}-T^{(q)}) = 0; \quad (5.13)$$

$c^{(q)}$ is specific heat at constant pressure.

† Due to kinetic and potential energies about the center of mass of the (q) component of the mixture.

Care must be exercised in identifying the transport properties of component (q) in the mixture when formulating equations for a component of a mixture. It is seen that for small deviation from mean motion, small concentration gradients, the relation in eqn. (5.8) and analogous derivations give

$$\mu_m^{(q)} = (\rho^{(q)}/\rho_m)\mu_m, \quad \mu_{m2}^{(q)} = (\rho^{(q)}/\rho_m)\mu_{m2}, \quad \kappa_m^{(q)} = (\rho^{(q)}c^{(q)}/\rho_m c_m)\kappa_m,$$

as approximations.[1]

It is seen that direct determination of $\mu_m^{(q)}$ according to kinetic theory cannot be made without a knowledge of the motion of the other components. However, when applied to a dilute suspension, simplifying approximations can be made.

5.4. Diffusion equation

Especially when dealing with cases with relaxation of a trace component in a mixture in the presence of an external field, it is more convenient to replace the continuity equation by a diffusion equation.

Rearranging eqn. (5.3) such that[73]

$$\frac{\partial \rho^{(q)}}{\partial t} = -\frac{\partial}{\partial x_j}(\rho^{(q)}U_{mj}) + \frac{\partial}{\partial x_j}[\rho^{(q)}(U_m - U_j^{(q)})] + \Gamma^{(q)}. \tag{5.14}$$

Since the flux of component (q) which may arise by diffusion and drift is given by

$$J_j^{(q)} = \rho^{(q)}(U_j^{(q)} - U_{mj}) \tag{5.15}$$

and the mole fraction $c^{(q)} = \rho^{(q)}/\rho_m$, we get

$$\rho_m \frac{dc^{(q)}}{dt} = -\frac{\partial J_j^{(q)}}{\partial x_j} + \Gamma^{(q)}. \tag{5.16}$$

For diffusion under field forces, $J_j^{(q)}$ is given by

$$J_j^{(q)} = J_{jD}^{(q)} + J_{jF}^{(q)}, \tag{5.17}$$

where $J_{jD}^{(q)}$ denotes diffusion, and $J_{jF}^{(q)}$ denotes flux due to field forces, and

$$J_{jD}^{(q)} = -D_m^{(q)} \frac{\partial \rho^{(q)}}{\partial x_j}, \tag{5.18}$$

where $J_{jF}^{(q)}$ is given by eqn. (5.5) and $D_m^{(q)}$ is the diffusivity of the (q) component in the mixture.

CHAPTER 6

DYNAMIC EQUATIONS OF A DILUTE SUSPENSION

With the knowledge of transport phenomena and parameters, we are ready to apply the equations of continuum mechanics to the general formulation of an aerosol suspension.

It is noted that continuum mechanics is directly applicable to molecular systems. When treating a particulate suspension, velocity and temperature of the fluid phase are mean values away from the particle–fluid interface; distributions of velocity and temperature around each particle are accounted for in terms of transport parameters.[47]

Any generalized formulation of an aerosol suspension has as its basis the recognized physical concepts concerning a single particle and a cloud of particles with or without size distribution. It must account for the following facts.

(1) The drag force acting on an aerosol particle is mainly due to the difference between the velocities of the particles and the stream. It is expected that the minimum size of an aerosol particle may consist of millions of molecules each (even for particles in the submicron range). Hence the random velocity of each particle due to its own thermal state is extremely low and the particulate phase does not contribute to the static pressure of the system (section 3.1).

(2) When a large number of aerosol particles is present, some average properties of a cloud of similar particles can be spoken of.

(3) Interactions exist between the particles and the stream, the solid particles themselves, and both with a given boundary. Heat transfer between the particles and the gas is due to the mean temperature difference. Fluctuation is expected, but each particle, due to its small size and high thermal conductivity (as compared to the gas), is assumed to be at a uniform temperature (section 3.2).

(4) Due to the difference in inertia of particles and fluid, one has to identify the velocity of the particulate phase and that of the gas separately at any given location in space. At a solid boundary, particles may have finite velocity even though the gas phase attains zero velocity there except at very low pressures.

(5) In a dilute suspension, due to dissipation in the wakes of particles, slowing down of particles by the fluid does not give rise to recovery of its velocity to static pressure of the mixture.[1,95] This constitutes a significant qualification on the continuum hydrodynamic equations in the previous section.

(6) When heat transfer is involved, radiation to and among particles can be significant.

(7) In an electromagnetic field, particles of the same species must be identified by similar electric charge q and dipole moment p. We neglect the effects of electric and magnetic dipoles in the present treatment, but they should be checked when larger gradients exist in the fields.[1]

Recognizing that particles of different size range constitute different phases from the point of view of continuum mechanics, the previous basic formulations[93] are generalized into a qualified continuum hydrodynamic formulation.[1] Here we illustrate with the case of dilute suspensions ($\Phi < 0.05$) in which the particles are sufficiently numerous. For the sake of simplicity, we shall neglect the bulk viscosities ζ in our formulations. Note that $\zeta = 0$ for incompressible fluid and its suspension of rigid particles.

For the convenience of formulation, we shall illustrate with monodispersed suspension, extension to a suspension with distribution in particle size is obvious from Chapter 5 and reference 1. For our present examples, we shall also neglect the generation of species.

6.1. Continuity equations

We define the density of a cloud of particles (of given size, material, charge, etc.) as ρ_p (the density of the solid material is denoted as $\bar{\rho}_p$)

$$\rho_p = n_p m, \tag{6.1}$$

where n_p is the number of particles per unit volume and m is the mass of each particle. The density of the gaseous phase in the mixture is

$$\rho = \bar{\rho}[1-(\rho_p/\bar{\rho}_p)]. \tag{6.2}$$

Note that $\rho_p/\bar{\rho}_p = \Phi$ is the volume fraction particles. Hence, for a dilute suspension $\rho \sim \bar{\rho}$. The overall density and mean velocity of the suspension is given by

$$\rho_m = \rho + \rho_p, \tag{6.3}$$

$$U_{mi} = \rho_m^{-1}[\rho U_i + \rho_p U_{pi}]. \tag{6.4}$$

In such a general system, these particles, in spite of visual discreteness, may be treated as a quasi-continuum, and the continuity relation in eqn. (5.3) gives

$$\frac{\partial \rho_p}{\partial t} + \frac{\partial \rho_p U_{pj}}{\partial x_j} = 0, \tag{6.5}$$

where U_{pj} is the jth component of the velocity of the cloud of particles. For the gaseous or fluid phase, the continuity equation is

$$\frac{\partial \rho}{\partial t} + \frac{\partial \rho U_j}{\partial x_j} = 0, \tag{6.6}$$

where ρ is the density of the gas in the suspension, and U_j is the jth component of velocity of the fluid phase with the volume occupied by the particulate phase excluded.

6.2. Momentum equations

Dissipation in the wakes of the particles are accounted for in the equation of momentum of the whole mixture of the gas and the particle phase. For a dilute suspension, viscous forces are contributed by the principal fluid, transfer of momentum of particles takes place by their diffusion, or $\mu_{mf} \sim \bar{\mu}$, $\mu_{mp} \sim \rho_p D_p$; D_p arises due to Brownian motion, interaction of wakes of particles, or turbulence in the fluid. The effectiveness of momentum transfer from particles to gas is accounted for by a parameter K, $K = 1$ for an accelerating (in speed) system and $K = 0$ for a decelerating system.[93,95] We have

$$\rho\left(\frac{dU_i}{dt}\right)+K\rho_p\frac{dU_{pi}}{dt_p}=-\left(\frac{\partial P}{\partial x_i}\right)+\frac{\partial}{\partial x_j}[\bar{\mu}\Delta_{ji}]+K\frac{\partial}{\partial x_j}[\rho_pD_p(\Delta_p)_{ji}]+\rho\boldsymbol{F}_i+K\rho_p\boldsymbol{f}_{pi}/m, \quad (6.7)$$

where $\boldsymbol{f}_{pi}$ is the force acting on each particle by the external field.

The momentum equation of the particulate phase is now

$$\rho_p\frac{dU_{pi}}{dt_p}=-\left(\frac{\rho_p}{\bar{\rho}_p}\right)\frac{\partial P}{\partial x_j}+\rho_pF(U_i-U_{pi})+\frac{\partial}{\partial x_j}[\rho_pD_p(\Delta_p)_{ji}]+\rho_p(\boldsymbol{f}_{pi}/m)+\rho_p(\boldsymbol{f}'_{pi}/m), \quad (6.8)$$

when $\boldsymbol{f}'_{pi}$ is the fluid force acting on each particle given in eqn. (3.15).

The momentum equation of the fluid phase alone is given by subtracting eqn. (6.8) from eqn. (6.7):

$$\rho\left(\frac{dU_i}{dt}\right)=-\left(1-K\frac{\rho_p}{\bar{\rho}_p}\right)\frac{\partial P}{\partial x_i}+\frac{\partial}{\partial x_j}(\bar{\mu}\Delta_{ji})-KF\rho_p(U_i-U_{pi})+\rho\boldsymbol{F}_i-K\rho_p\boldsymbol{f}'_{pi}/m. \quad (6.9)$$

For a dilute suspension, therefore, motion of the fluid phase is unaffected by the presence of the particles. It is noted that of eqns. (6.7) to (6.9), only two are independent.

6.3. Energy equations

When applied to a dilute suspension, the thermal conductivity of the fluid phase in the mixture $\kappa_{mf}\sim\bar{\kappa}$, and that of the particulate phase $\kappa_{mp}\sim c_p\rho_pD_p$. The energy equations of the phases are:

$$\rho\frac{d}{dt}\left(\frac{U^2}{2}+cT\right)+\rho_p\frac{d}{dt_p}\left(\frac{U_p^2}{2}\right)=\frac{\partial P}{\partial t}+\frac{\partial}{\partial x_j}(\bar{\mu}U_j\Delta_{ji})+\frac{\partial}{\partial x_j}\left(\bar{\kappa}\frac{\partial T}{\partial x_j}\right)-$$

$$-c_p\rho_pG(T-T_p)+\rho F(U_j-U_{pj})^2+\boldsymbol{G}, \quad (6.10)$$

$$\frac{dT_p}{dt_p}=G(T-T_p)+\frac{g_p}{c_p}+(c_p\rho_p)^{-1}\frac{\partial}{\partial x_j}\left[c_p\rho_pD_p\left(\frac{\partial T_p}{\partial x_j}\right)\right], \quad (6.11)$$

where c is the specific heat at constant pressure of the fluid, $\boldsymbol{G}$ is the rate of heat generated per unit volume of the fluid including radiative input, and g_p is the heat input to each particle. Equation (6.10) accounts for the fact that U_p arises due to U and eqn. (6.11) speaks of the fact that heat added to the particle cloud raises the body temperature of the particles.

6.4. Diffusion under field forces

For a non-reactive dilute suspension, eqn. (5.16) reduces to

$$\frac{d\rho_p}{dt}=-\frac{\partial J_j}{\partial x_j} \quad (6.12)$$

with

$$J_j=-\left(D_p\frac{\partial\rho_p}{\partial x_j}\right)+[\rho_p(U_{pj}-U_j)], \quad (6.13)$$

and the last term is given by the relations in eqn. (6.8). When

$$\text{field forces} > \text{fluid forces} > \text{inertia forces} \quad (6.14)$$

in a dilute suspension

$$\rho_p(U_{pj}-U_{ji})=\frac{\rho_p}{F}(\boldsymbol{f}_{pj}+\boldsymbol{f}'_{pj})/m. \quad (6.15)$$

This is also true for fully developed motion even if the condition in eqn. (6.14) does not apply.

Now we are ready to discuss a number of cases reported recently. We shall not deal with cases of heat transfer which have been covered in a recent review.[73]

CHAPTER 7

ONE-DIMENSIONAL MOTION OF CHARGED CLOUD OF PARTICLES

Flow of a suspension with interaction of electric, gravitational and hydrodynamic fields occurs in dust settling and precipitation, cloud charges, and thunderstorm phenomena. Simplified and pertinent examples will be illustrated here.

Performance of electrostatic precipitators in the aspects of charging field configuration and distribution are presented in the treatise by White.[(14)] With the above formulations, fundamental aspects of electrohydrodynamic flow can be explored. Identification of similarity relations in this type of flow will serve as reference to model studies and scaling.

Elementary formulation of flow of a gas with space charge (electrohydrodynamics) was given in studies by Stuetzer.[(96)] Space charge flow with various boundary configurations was made by Lockwood and Hamza,[(97)] extending from earlier efforts of Langmuir and Blodgett.[(98)] Motion of charged aerosol particles in relation to surfaces has also been utilized as basic methods for analyzing particle size[(75,99)] as well as electrostatic precipitation.[(37)] In all of these studies, theoretical analysis includes that of single particles under viscous drag, motion of microscopic particles in an electrostatic field, or numerical summation of discrete particles in hydrodynamic potential flow field. In studies by Soo,[(93,94)] electrohydrodynamic flow of a cloud of charged solid particles suspended in a gas was illustrated. It was shown that the effect of electric charge on aerosols, even for very small charge to mass ratio, could be significant. The effect is felt even when a short flow length is involved, such as in a nozzle handling gas–solid flow.[(95)]

In this chapter, cases of one-dimensional motion are illustrated.

7.1. Settling of charged dusts

Extending the case of two-dimensional steady motion,[(1,93)] we shall now consider the interaction of gravitation and electric fields in the settling of charged aerosols. The simplified system is defined in Fig. 6, which shows that a uniform non-reactive gaseous suspension of charged aerosols of uniform size (radius a, mass m), charge q, and uniform cloud density ρ_{p0} is suddenly brought between two grounded conductor plates of infinite extent at distance y_0 apart normal to the direction of gravity of acceleration g. Obviously, settling will take place at the surface of both the upper and lower plates, but at different rates. Changes in this system with time are computed as follows. When applied to this one-dimensional unsteady system, eqns. (6.5) and (6.8) give

$$(\partial \rho_p/\partial t)+(\partial \rho_p v_p/\partial y) = 0, \tag{7.1}$$

$$(\partial v_p/\partial t)+v_p(\partial v_p/\partial y) = -g+(q/m)E_y+F(-v_p), \tag{7.2}$$

where v_p is the velocity. The electric field E_y is given by the Poisson equation

$$\partial E_y/\partial y = \varepsilon_0^{-1}(q/m)\rho_p. \tag{7.3}$$

The initial conditions are:

$$t=0, \quad \rho_p = \rho_{po}, \quad v_p = 0, \quad dv_p/dt = -g. \tag{7.4}$$

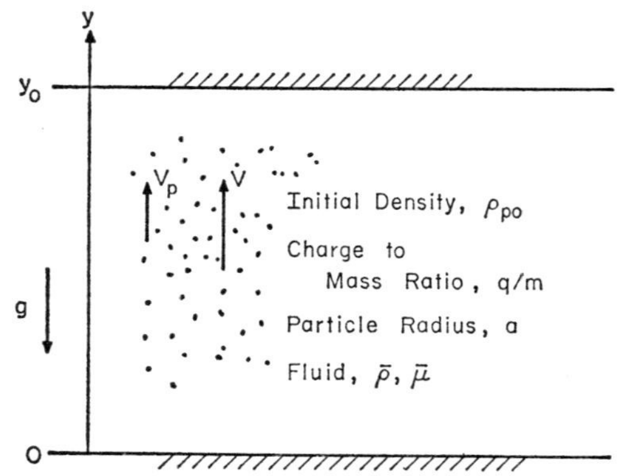

FIG. 6. Settling of a uniformly charged dust suspension between two grounded plates at distance y_0 apart.

The solution of these equations is facilitated by applying the Mises transformation with the introduction of "stream function" ψ_p such that

$$\rho_p = -\partial\psi_p/\partial y, \qquad \rho_p v_p = \partial\psi_p/\partial t, \tag{7.5}$$

whereby eqn. (7.1) is satisfied. Changing the independent variables from y and t to ψ_p and t, eqns. (7.2) and (7.3) become

$$(\partial/\partial t)[v_p \exp(Ft)] = -g \exp(Ft)+(q/m)E_y \exp(Ft) \tag{7.6}$$

and

$$(\partial E_y/\partial\psi_p) = -\varepsilon_0^{-1}(q/m). \tag{7.7}$$

Elimination of E_y between eqns. (7.6) and (7.7) gives

$$(\partial^2/\partial\psi_p\partial t)[v_p \exp(Ft)] = -\varepsilon_0^{-1}(q/m)^2 \exp(Ft) \tag{7.8}$$

with the initial condition changed to

$$t=0, \quad \psi_p = \rho_{p0}\left(\frac{y_0}{2}-y\right), \quad v_p = 0, \quad (\partial v_p/\partial t)_\psi = -g. \tag{7.9}$$

Solution of eqn. (7.8) subjecting to eqn. (7.9) gives

$$-v_p = [(q/m)^2/\varepsilon_0 F]\psi_p[1-\exp(-Ft)]+gt\exp(-Ft) \tag{7.10}$$

or, in dimensionless form, with $Ft = t^*$, $(\psi_p/\rho_{p0}y_0) = \psi_p^*$,

$$-(v_p/Fy_0) = -v_p^* = (N_{ev})[1-\exp(-t^*)]\psi_p^*+(N_m)_g t^* \exp(-t^*), \tag{7.11}$$

with dimensionless quantities, the electroviscous number

$$(N_{ev}) = \rho_{p0}(q/m)^2/\varepsilon_0 F^2, \tag{7.12}$$

and the gas–particle momentum transfer number in the gravitational field

$$(N_m)_g = g/y_0 F^2. \tag{7.13}$$

The predominance of the influence of either the electrostatic effect or the gravity effect is seen as the relative magnitude of (N_{ev}) and $(N_m)_g$, or whether the ratio

$$(N_{ev})/(N_m)_g = \rho_{p0}(q/m)^2 y_0/\varepsilon_0 g \tag{7.14}$$

is much greater or much smaller than 1. It is seen that for $\rho_{p0} = 10^{-2}$ kg/m³, $(q/m) = 10^{-4}$ coul/kg, $y_0 \sim 1$ m, $(N_{ev})/(N_m)_g \sim 1$ or that both gravity effect and electrostatic effect have equal significance in settling.

To determine ρ_p and ψ_p, we substitute eqn. (7.11) with eqn. (7.5) into eqn. (7.1) to get, for $\rho_p^* = \rho_p/\rho_{p0}$,

$$(\partial\rho_p^*/\partial t^*)+(N_{ev})[1-\exp(-t^*)](\partial/\partial y^*)(\rho_p^* \int \rho_p^* \, dy^*)$$
$$-(N_m)_g t^* \exp(-t^*)(\partial\rho_p^*/\partial y^*) = 0 \tag{7.15}$$

with $y^* = y/y_0$. In eqn. (7.15), the second term gives the change in density due to space charge effect, and the third term gives the change in density due to image charge and gravity effects. Equation (7.15) is not readily solved in analytical form, but some understanding of the problem can be obtained by considering a few limiting conditions.

(a) When the particles are not charged, $(q/m) = 0$, eqn. (7.15) takes the form

$$\partial\rho_p^*/\partial t^* = (N_m)_g t^* \exp(-t^*)(\partial\rho_p^*/\partial y^*). \tag{7.16}$$

Its pertinent solution for the given conditions is

$$\rho_p^* = 1 \ at \ y^* \leq 1-(N_m)_g[1-(1+t^*)\exp(-t^*)], \tag{7.17}$$
$$\rho_p^* = 0 \ at \ y^* > 1-(N_m)_g[1-(1+t^*)\exp(-t^*)]. \tag{7.18}$$

Noting that at small t^*, eqn. (7.17) simplifies to

$$y^* \leq 1-(N_m)_g t^{*2}, \tag{7.19}$$

which accounts for the starting of the descent. This compares to:

$$y^* \leq 1-(N_m)_g t^* \tag{7.20}$$

or

$$y \leq y_0-(g/F)t, \tag{7.21}$$

in the case of long time batch settling given by Soo.[1] It is noted that for aerosols of 1 μ diameter in the atmosphere, $F \sim 10^5$, therefore in 1 sec $t^* = 10^5$.

(b) For $N_{ev} \gg (N_m)_g$ at sufficiently large t^*, eqn. (7.15) reverts to

$$(\partial\rho_p^*/\partial t_1^*)+(\partial/\partial y^*)(\rho_p^* \int \rho_p^* \, dy) = 0; \tag{7.22}$$

a similar form for the case of steady electrohydrodynamic flow into a channel given by Soo[1,93] with

$$t_1^* = [t^*+\exp(-t^*)](N_{ev}). \tag{7.23}$$

The solution now takes the form

$$\rho_p^* = (1+t_1^*)^{-1}+(y^*-\tfrac{1}{2})^2(1+t_1^*)^{-4}+\ldots, \tag{7.24}$$
$$\psi_p^* = -(y^*-\tfrac{1}{2})(1+t_1^*)^{-1}-(1/3)(y^*-\tfrac{1}{2})^3(1+t_1^*)^{-4}+\ldots. \tag{7.25}$$

and

$$v_p^* = -(N_{ev})[1-\exp(-t_1^*)]\psi_p^* \tag{7.26}$$

as given in Fig. 7. The E field is given by eqn. (7.7) and

$$E^* = \varepsilon_0 E/(q/m)\rho_{p0} y_0 \psi_p^*. \tag{7.27}$$

(c) When $N_{ev} \sim (N_m)_g$, both the electrostatic effect and gravity effect are prominent. A complete solution of eqn. (7.15) must embody the above features, namely deposition toward both the top and the bottom plates due to electrostatic effect, with less toward the top plate than the bottom plate due to gravity effect. The orders of magnitude are seen in that for $\rho_{p0} = 1$ kg/m^3, $q/m = 10^{-4}$ coul/kg, $a = 5$ μ at room condition, $F \sim 3000$, $N_{ev} \sim 10^{-4}$, and $(N_m)_g \sim 10^{-6}$ for $y_0 = 1$ m.

Experimental work on settling of dust at both top and bottom of horizontal plates was reported by Owen,(100) showing more collection at the bottom plate than at the top plate, as suggested by the above example.

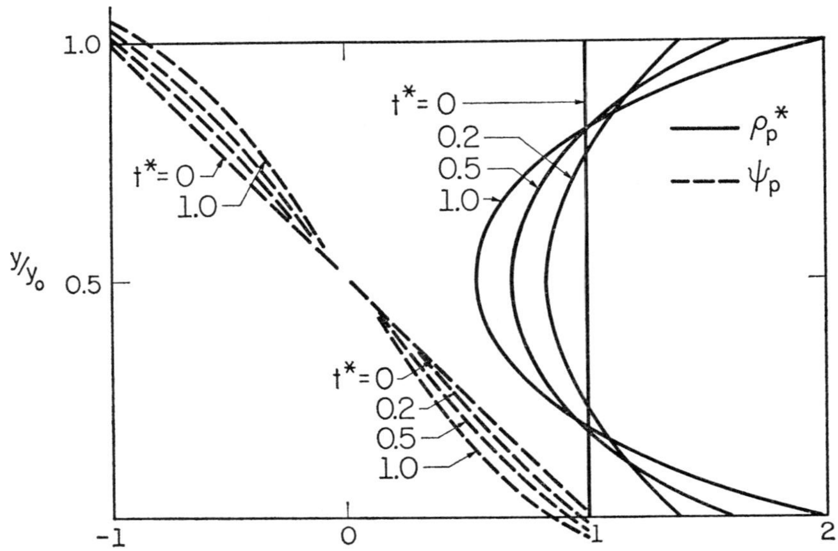

FIG. 7. Behavior settling between two walls, $N_{ev} \gg (N_m)_g$.

Application of the above method of solution to the case of an electrostatic precipitator passage was illustrated by Soo,(93) showing more collection at the inlet than further downstream.

The effect of an alternating current electric field on the channel flow of a suspension was studied by Soo(73) showing that maximum heat transfer rate occurs when the frequency of the electric field gives the greatest number of impact of particles with the walls per unit time.

7.2. Electric field in a rain cloud

Cloud, fog, and rain come under the definition of dilute particulate suspension. A severe thunderstorm of 40 mm/hr at 6 m/sec speed of fall amounts to a density ρ_p of nearly 2×10^{-3} kg/m^3.

To explain the phenomenon of thunderstorms, Shishkin(101) computed the electric field due to charges on cloud particles and rain drops.

First, he considered the case of steady settling of charged cloud particles without coagulation. The electric field is uniform in this case due to compensation of convection current produced by falling drops by the conduction current in the electric field via separation of space charges. In one dimension,

$$-(q/m)\rho_p v_p = \sigma E_y, \tag{7.28}$$

where σ is the electrical conductivity. Equation (7.2) is now reduced to

$$Fv_p = -g+(q/m)E_y. \tag{7.29}$$

Hence the electric field E_y is given by:

$$E_y = g\{(q/m)+[F\sigma/(q/m)\rho_p]\}^{-1}. \tag{7.30}$$

For $\rho_p = 10^{-3}$ kg/m³, $a = 10\ \mu$, $\bar{\mu} = 1.74\times10^{-5}$ kg/m sec, $F = 9\ \bar{\mu}/2\rho_p a^2 = 783\ \text{sec}^{-1}$, q given by eqn. (2.31) with $\varphi = (1/3)\times10^{-11}$ coul/m, or $(q/m) = (1/4\pi)\times10^{-4}$ coul/kg, $\sigma = 4\times10^{-14}$ mho/m, $g = 9.8$ m/sec², we get $E_y \sim 10^3$ v/m, a rather moderate field. Fine weather field at the surface of the earth is 120 v/m.(3) It is noted that in the present case, $N_{ev} \sim 2\times10^{-8}$, and for a height of 30 m, $(N_m)_g \sim 5\times10^{-7}$.

7.3. Effect of charge on growing rain drops

Application of the dynamic equations to a cloud of changing size distribution has not been made. However, the technique of handling a cloud of uniformly changing particle size is illustrated in an example given by Shishkin.(101) Shishkin took the example where increase of droplet size of species 1 by coalescence of small droplets of species 2 occurs over a stratum. This results in enormous accumulation of charges according to the specific charge q_1 given by eqn. (2.33). As a rough approximation, the contribution of charges on background droplets of species 2 such as in the above example and of air ions to this accumulated space charge is neglected. This leads to

$$dE_y/dy = \varepsilon_0^{-1} q n_{p1}. \tag{7.31}$$

It is seen that as long as uniform growth of a given number of particles of uniform size is dealt with, there is no need for treating size change as phase change.(1) With the increase in a_1 as given in eqn. (4.28), together with eqn. (2.33), eqn. (7.31) gives

$$dE_y/da_1 = -(4\pi/3\varepsilon_0)(\bar{\rho}_p/\rho_{p2}\eta_{12}\eta_{co})n_p^{(s)} q_1 a_1^3, \tag{7.32}$$

which integrates to

$$E_y = -(\pi/3\varepsilon_0)(\bar{\rho}_p/\rho_{p2}\eta_{12}\eta_{co})n_{p1} q_1 a_1^4, \tag{7.33}$$

which for $q_1 = (1/300)$ coul/m³ (or (10 e.s.u./cm³), $\rho_{p2} = 10^{-3}$ kg/m³, $a_1 = 1$ mm, rain intensity of 10 mm/hr (1/360 kg/m² sec) at 6 m/sec. For the fall velocity given by eqn. (4.34), $n_{p1} \sim 100\ \text{m}^{-3}$. Taking $\eta_{12}\eta_{c0} = 1/4$, we get $E \sim 2\times10^5$ v/m. Over a height of 30 m of a region of 30 km², a thunderstorm charge of nearly 1.5 coul can be accumulated. In this case, $(N_{ev}) \sim 1$ and $(N_m)_g \sim 5\times10^2$.

If lightning occurs with a discharge of the above magnitude, Shishkin gave an estimate of one discharge every 6 sec. It is noted that the average discharge of a lightning flash is to the order of 20–30 coul.(3) From a study of precipitation and lightning from convective clouds, Battan confirmed this magnitude and gave a charge to mass ratio of 3.3×10^{-7} coul/kg.(102)

Various observations on thunderstorms, their electrification by impacts with ice crystals and generation of rain up to 100 mm/hr were discussed.(103–105) Variations of drop size and transients have made definite correlations difficult.

7.4. Effect of diffusion on fall-out

In the above section, we have neglected the effect of diffusion on the motion of a charged suspension. Brownian diffusion is significant to the settling of charged aerosols below 1 μ size in the atmosphere; turbulent diffusion is important to the transport of aerosols in the atmosphere or in a flow system.

Some particles in atmospheric fall-out may include nuclear dusts, micrometeorites, cosmic dusts, and industrial pollutants. Aerosols of sulfur compounds have been detected at 20 km above the earth's surface. Recent studies on the descent of small particles through a stagnant isothermal atmosphere (average molecular diameter 1.7 Å), such as those reported by Pasquill[106] and the study by Banister and Davis,[107] account for both gravitational fall and molecular diffusion, but not the effect of electric charge on the particles. Banister and Davis dealt with particles 4–10 Å in diameter with $\bar{\rho}_p = 6\times10^3$ kg/m^3 and formulated the problem with these assumptions: (1) The atmosphere is stagnant and isothermal (vertical turbulence between 15–50 km is known to be negligible). (2) The suspension is dilute. (3) The initial injection is widely spread such that the problem may be treated as one-dimensional. Their solution is readily applicable to the case of the charged contaminant particles if we further make the assumption that: (4) The electric field is uniform due to balance of convection current of falling particles by the conduction current in the electric field by charge separation eqn. (7.28). The fine-weather atmosphere[3] has a conductivity nearly 2.5×10^{-4} mho/m.

For this case, eqns. (6.8), (6.12) to (6.15) give[107]

$$\frac{\partial \rho_p}{\partial t} = \frac{\partial}{\partial z}\left[D_p \frac{\partial \rho_p}{\partial z} + \frac{\rho_p}{F}\frac{f_p}{m_p}\right], \tag{7.34}$$

where z is the altitude with an upward positive direction, and for particle mass m_p and molecular mass m_a,

$$D_p = \frac{3}{8}\left[\frac{\pi(m_a+m_p)kT}{2m_a m_p}\right]^{1/2} \Big/ n_a\pi(a_a+a)^2, \tag{7.35}$$

where n is the number density of air molecules, a_a is its radius, a is that of the particles, k is the Boltzmann constant, T is the temperature, and the drift velocity is given by

$$\frac{f_p}{Fm_p} = \frac{3}{8}\left[g+E_z\left(\frac{q}{m_p}\right)\right]\left[\frac{m_a m_p(m_a+m_p)}{2\pi kT}\right]^{1/2} \Big/ (a_a+a)^2, \tag{7.36}$$

when the effect of electric field of the atmosphere is added. Note that $\rho = n_a m_a$ is the density of air.

Numerical solution of eqn. (7.34) was carried out by Banister and Davis for a number of situations at $T = 245°$K and $\rho_0 = 1.72\times10^{-2}$ kg/m^3 at ground level of an isothermal experimental atmosphere and aerosol particles of density of 6×10^3 kg/m^3. For parti cles of corresponding molecular weight of 150 and 1500, the parameters chosen were:

Particle molecular weight	150	1500
Particle radius, a (Å)	1.7	4.65
ρD_p (kg/m^3)(m^2/sec)	1.4×10^{-5}	0.48×10^{-5}
$\rho f_p/Fm_p$ (kg/m^3)(m/sec)	0.829×10^{-10}	3.05×10^{-10}

The results are given in Fig. 8, showing the relative contribution of diffusion and drift of the above two cases, starting with square distribution of injected contaminant. Both cases show the tendency of descending rapidly to a certain altitude where the spatial concentration profile becomes constant thereafter.

It is seen that as long as the electric field remains constant, the methods of Banister and Davis[107] are readily adapted. Where E is uniform, both particle charge and E field tend to be small. $E = 10^3$ v/m and $q/m = 10^{-4}$ contributes to a slight modification of the drift

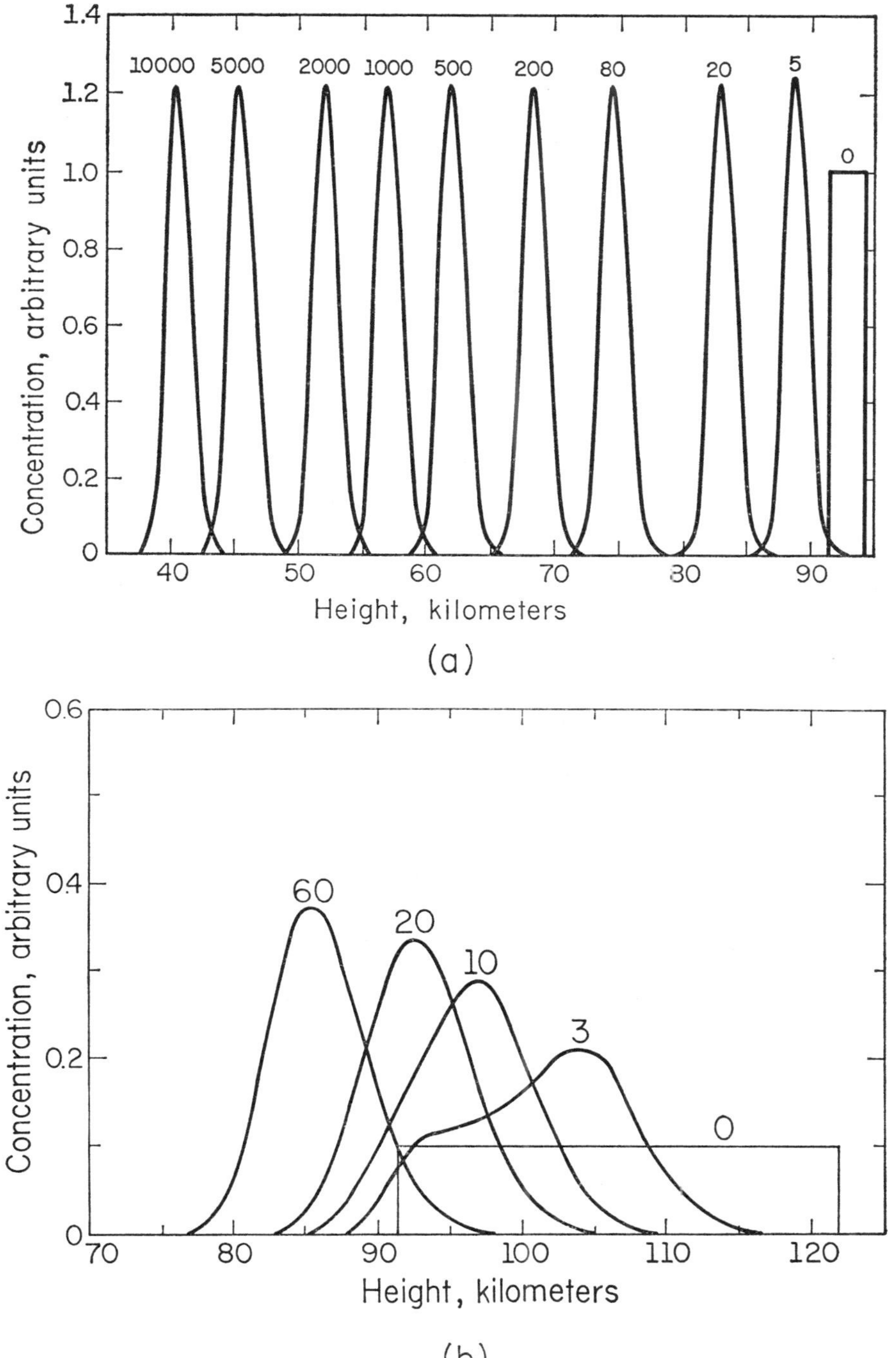

FIG. 8. Time history of contaminant profiles.[107] (a) Injection between 91.44 and 94.44 km of particles of equivalent molecular weight of 1500—large drift, low diffusivity. (b) Injection between 91.44 and 121.92 km of particles of equivalent molecular weight of 150—slow drift, high diffusivity.

parameter in eqn. (7.36). However, an interesting situation would be where space charge is prominent, the field is non-uniform, and particles possess large charge to mass ratio. These effects modify the nature of eqn. (7.34) greatly and remain to be studied.

Other cases of interest are large particles (2–6 mm diameter) in a turbulent field[108] and significant particle diffusivity.[109] An analytical solution is available for the condition of constant mean density ρ and is limited to low altitude release of contaminant such as in cases of air pollution.[110] This analytical solution to the three-dimensional transient diffusion equation is applicable to the release of a contaminant of species k of total mass m_k at a point represented by the coordinates $x = y = 0$, $x = h$ above ground at time $t = 0$. The solution was obtained for constant density of atmosphere ρ, diffusivity of species k: D D_{kx}, D_{ky}, and D_{kz} being constant or being a function of t. The density distribution takes the form

$$\rho_k = \frac{m_k}{\pi^{3/2}h^3}\frac{D_{kz}}{(D_{kx}D_{ky})^{1/2}}\exp\left[-\frac{x^2}{4D_{kx}t}-\frac{y^2}{4D_{ky}t}\right]\times$$
$$\times t_k^{*-3/2}\left\{\exp\left[-\frac{1}{4t_k^*}\left(\frac{z}{h}+\tau_k^*t_k^*-1\right)^2\right]+\right.$$
$$\left.+\alpha_k\exp\left[-\frac{1}{4t_k^*}\left(\frac{z}{h}+\tau_k^*t_k^*+1\right)^2\right]\right\}, \tag{7.37}$$

with $t_k^* = D_{kz}t/h^2$, $\tau_k^* = gh/D_{kz}F_k$, and $\tau_k^*t_k^* = gt/hF_k$. The case of constant electric field E_z for charge to mass ratio $(q/m)_k$ is accounted for by replacing g by $[g+E_z(q/m)_k]$ as before. α_k is the ground reflection coefficient; $\alpha_k = 1$ for a non-absorbing ground and $\alpha_k = 0$ for a completely absorbing ground. The dimensionless ground level flux is then

$$J_{kG}^{\prime *} = J_{kG}'\frac{\pi^{3/2}h^4}{m_k D_{kz}}\frac{(D_{kx}D_{ky})^{1/2}}{D_{kz}}\exp\left[\frac{x^2}{4D_{kx}t}+\frac{y^2}{4D_{ky}t}\right]$$
$$= \frac{1}{2}\left(\frac{\tau_k^*t_k^*+1}{t_k^{*5/2}}\right)\exp\left[-\frac{(\tau_k^*t_k^*-1)^2}{4t_k^*}\right]+\frac{\alpha_k}{2}\left(\frac{\tau_k^*t_k^*-1}{t_k^{*5/2}}\right)\exp\left[-\frac{(\tau_k^*t_k^*+1)^2}{4t_k^*}\right]$$
$$= J_{kGF}'+J_{kGR}', \tag{7.38}$$

giving the variation of maximum ground level flux with time as $\tau_k^* \to 0$, the maximum of J_{kGF}', and minimum of $J_{kGR}^{\prime *}$ occur at $t_k^* = 1/10$. Curves were plotted with $J_{GF}^{\prime *}$ vs. t^* parametric with τ^*. For $\tau^* = 100$, the fall-out begins to be felt at nearly $t^* = 0.003$ after release, reaching a peak at $t^* = 0.01$, and ends at $t^* = 0.02$. At $\tau^* \to 0$, and absorbing ground ($\alpha_k = 0$), ground level flux reaches a maximum at $t^* = 1/10$ after release due to diffusion alone.

CHAPTER 8

FULLY DEVELOPED MOTION

Besides the applications to practical flow systems, fully developed flow in a circular pipe has been a favorite configuration for studying the effect of a solid boundary in fluid mechanics. When dealing with the flow of charged suspensions, fully developed pipe flow is again a suitable model for determining interactions among fluid, particle, fields, and wall. Knowledge under this topic developed from earlier semi-empirical studies(111) followed by rigorous solution for negligible gravitational effect(112) to a recent study by Soo accounting for interactions with simultaneous effects of gravitational and electric fields. His recent study applies to all possible orientations of the pipe with respect to gravity.

8.1. Basic relations of pipe flow

The equations of general motion of a suspension are reduced to simple forms when applied to a fully developed pipe flow. For radial, axial, and azimuthal coordinates r, z, φ; velocity components v, u, w for the fluid; and v_p, u_p, w_p for the particulate phase, fully developed flow exists with $\partial/\partial t = 0$, $v = v_p = 0$, $w = w_p = 0$, $\partial u/\partial z = \partial u_p/\partial z = 0$; t denotes time. For a pipe making an angle θ with the direction of gravity (Fig. 9), the overall momentum equation based on eqn. (6.7) in the z-direction for a single species of particles reduces to

$$-\frac{\partial P}{\partial z} - (\rho+\rho_p)\, g\cos\theta + \frac{1}{r}\frac{\partial}{\partial r} r(\tau_{zr}+\tau_{pzr}) + \frac{1}{r}\frac{\partial}{\partial \varphi}(\tau_{z\varphi}+\tau_{pz\varphi}) = 0, \tag{8.1}$$

where P is the static pressure, ρ, ρ_p are densities of the fluid phase and the particle cloud respectively, g is the gravitational acceleration, and τ, τ_p are shear stresses of the fluid and the particulate phases respectively.

Similar simplifications give, from eqn. (6.8), the equation of motion of the particulate phase of spheres:

$$-\rho_p\left(1-\frac{\bar{\rho}}{\bar{\rho}_p}\right) g\cos\theta + \frac{1}{r}\frac{\partial}{\partial r} r\tau_{pzr} + \frac{1}{r}\frac{\partial}{\partial \varphi}\tau_{pz\varphi} + F\rho_p(u-u_p) = 0, \tag{8.2}$$

where F is the inverse of relaxation time for momentum transfer from the fluid and the particle including the correction for apparent mass in eqn. (3.15),

$$F = F^*(9\bar{\mu}/2a^2\bar{\rho}_p)/[1+(\bar{\rho}/2\bar{\rho}_p)] \tag{8.3}$$

with $F^* = (C_D/24)\,(2a\bar{\rho}|u-u_p|/\bar{\mu})$, C_D being the drag coefficient of a sphere; $F^* = 1$ in the range when Stokes law of drag of a sphere applies. In addition, $\bar{\mu}$, $\bar{\rho}$ are the viscosity and the density of the material constituting the fluid phase, $\bar{\rho}_p$ is the density of material constituting the particulate phase, a is the radius of the particles.

Fully-developed motion excludes deposition of the particulate phase under field force $\boldsymbol{f}_p$; that is, the fluxes due to diffusion and relaxation under field force $\mathbf{J}_F$ must be equalized;† eqn. (6.13) gives

$$\mathbf{J}_F + \left(-D_p \frac{\partial \rho_p}{\partial \mathbf{r}}\right) = 0, \tag{8.4}$$

where D_p is the diffusivity of the particle phase, $\mathbf{J}_F$ is determined by the eqn. (6.8) of particle motion. Since $d\mathbf{u}_p/dt_p = (\partial \mathbf{u}_p/dt) + \mathbf{u}_p \cdot (\partial \mathbf{u}_p/\partial \mathbf{r}) = 0$ for steady fully developed motion, we get

$$\mathbf{J}_F = \rho_p(\mathbf{u}_p - \mathbf{u}) = \rho_p \boldsymbol{f}_p / mF. \tag{8.5}$$

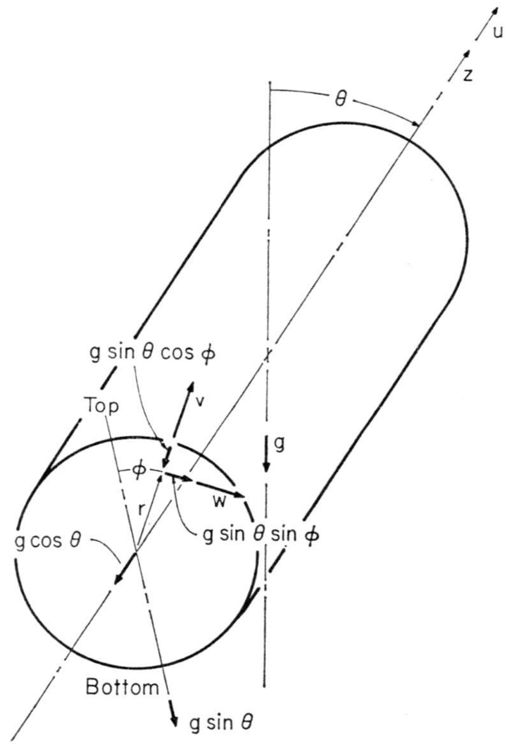

FIG. 9. Coordinate system and components of gravitational acceleration.

$\boldsymbol{f}_p$ includes the force acting on the particles due to gravity, the lift force acting on a particle in the radial direction due to fluid shear, the Magnus effect as given by eqn. (3.9):

$$f_L = 6.46 \bar{\rho}^{1/2} \bar{\mu}^{1/2} (u_p - u + \Delta u) \left| \frac{du}{dr} \right|^{1/2} a^2, \tag{8.6}$$

where Δu is the effective particle velocity due to finite radius a when $u_p = u$ locally but shear rate is finite. $\boldsymbol{f}_L$ is negligible in this case since the Magnus effect is usually limited to within the thickness of the laminar sublayer $\boldsymbol{f}_p$ also includes force due to electric field $\mathbf{E}$

† When deposition of particles is accounted for, the solution becomes analogous to the Graetz problem for heat transfer in single phase pipe flow.[132]

$$f_E = q\mathbf{E} = q(-\partial V/\partial \mathbf{r}) \tag{8.7}$$

for charge q and electric potential V of the particle cloud. Neglecting f_L for the present, it is readily shown that eqn. (8.4) integrates to,

$$D_p \ln \rho_p = C_1 - (q/m)\frac{1}{F} V - g\frac{\sin\theta r}{F}\cos\varphi\left(1-\frac{\bar{\rho}}{\bar{\rho}_p}\right), \tag{8.8}$$

and the constant of integration $C_1(\varphi) = C_1(r) = C_1$ is a constant. The lift force acting on the particles due to fluid shear is negligible when the thickness of the laminar sublayer of the fluid $y_s = 60(u_0 R\bar{\rho}/\bar{\mu})^{-7/8}R \leq a$; u_0 is the maximum velocity of the fluid and R is the pipe radius. For a steady, fully developed flow in an infinitely long pipe, V is given by the Poisson equation in the form

$$\frac{1}{r}\frac{\partial}{\partial r} r \frac{\partial V}{\partial r} + \frac{1}{r^2}\frac{\partial^2 V}{\partial \varphi^2} = -(q/m)\frac{\rho_p}{\varepsilon}, \tag{8.9}$$

where ε is the permittivity.

The condition of a dilute suspension $\rho \sim \bar{\rho}$ permits the stipulation that u, τ of the fluid are not influenced by the presence of the particles. Further, for large Froude numbers of the fluid phase $u_0/\sqrt{(2Rg)}$, u_0 being the maximum velocity of the fluid phase, R the pipe radius, gravity effect does not influence the velocity distribution of the fluid phase, $\tau = \tau_{zr}$ alone in eqn. (8.1). For turbulent flow, let $y = R-r$,

$$u = u_0(y/R)^{1/7} \tag{8.10}$$

for $y_s < y < R$. Under this condition, gravity effect contributes only to the static pressure.

For a dilute suspension, transport of momentum of the particulate phase is by diffusion of the particles through the fluid. When the particle–fluid interaction length L_p is not too small compared to R, we have, from eqn. (4.24)

$$\left.\begin{aligned} \tau_{pzr} &= \rho_p D_p \frac{\partial u_p}{\partial r}, \\ \tau_{pz\varphi} &= \rho_p D_p \frac{\partial u_p}{r\partial \varphi}, \end{aligned}\right\} \tag{8.11}$$

in eqn. (8.2).

With u as prescribed by eqn. (8.10), simultaneous solution of eqns. (8.2), (8.8), and (8.9) gives the velocity u_p, density ρ_p, and electric potential V of the particulate phase.

The boundary conditions are

$$V(R, \varphi) = V_R(\varphi), \tag{8.12}$$

and

$$u_{pw} = -L_p \left.\frac{du_p}{dr}\right|_{r=(R-a)} \tag{8.13}$$

for the slip velocity of particles along the wall, and

$$\int_0^{2\pi}\int_0^{R-a} \rho_p u_p r\, dr\, d\varphi = \dot{m}, \tag{8.14}$$

where $\dot{m}$ is the total flow rate of particles. For a conducting pipe, the surface at radius R is a Gaussian surface such that V_R is uniform.

It is readily seen from eqn. (8.8) that for a pipe of a conductor $V_R(\varphi)$ = constant and gravity effect gives the relation of particle density at the top $\rho_p(R, 0)$ and bottom $\rho_p(R, \pi)$ as

$$\ln\frac{\rho_p(R,\pi)}{\rho_p(R,0)} = \frac{2g\sin\theta R}{D_p F}\left(1-\frac{\bar{\rho}}{\bar{\rho}_p}\right). \tag{8.15}$$

Therefore the significance of the effect of gravity depends on this dimensionless quantity and the magnitude of $\rho_p(R, \pi)-\rho_p(R, 0)$ depends on actual density and θ.

8.2. Characteristic relations of pipe flow

Introducing the following dimensionless quantities into the above relation

$$\left.\begin{aligned} r^* &= r/R = 1-y^*, & \rho^* &= \rho_p/\rho_{p0}, \\ u^* &= u/u_0, \quad u_p^* = u_p/u_0, & V^* &= \frac{(q/m)V}{D_pF} \end{aligned}\right\} \tag{8.16}$$

where ρ^* is normalized by the following dimensionless total flow rate and ρ_{p0} is defined as the particle density at the center of the pipe:

$$\frac{\dot{m}}{\pi R^2 \rho_{p0} u_0} = \frac{1}{\pi}\int_0^{2\pi}\int_0^1 \rho^* u_p^* r^* \, dr^* \, d\varphi. \tag{8.17}$$

Equations (8.2) and (8.8) become, for given $u^* = y^{*1/7}$ for instance, with Reynolds number $N_{Re} = Ru_0\bar{\rho}/\bar{\mu}$ in the turbulent range,

$$-\tfrac{1}{2}\gamma\rho^* + \frac{1}{r^*}\frac{\partial}{\partial r^*} r^*\rho^* \frac{\partial u_p^*}{\partial r^*} + \frac{1}{r^{*2}}\frac{\partial}{\partial \varphi}\rho^*\frac{\partial u_p^*}{\partial \varphi} + \beta\rho^*(u^*-u_p^*) = 0, \tag{8.18}$$

$$\ln \rho^* = -V^* - \tfrac{1}{2}\eta r^* \cos\varphi, \tag{8.19}$$

$$\frac{1}{r^*}\frac{\partial}{\partial r^*} r^* \frac{\partial V^*}{\partial r^*} + \frac{1}{r^{*2}}\frac{\partial^2 V^*}{\partial \varphi^2} = -4\alpha\rho^*, \tag{8.20}$$

the constant of integration in eqn. (8.8) is set to be $C_1 = 0$, and the coefficients α, β, γ, and η are

$$\left.\begin{aligned} \alpha &= N_{ev}N_{DF} = \frac{\rho_{p0}}{4\varepsilon_0}(q/m)^2\frac{R^2}{D_pF}, \\ \beta &= N_{DF}^{-1} = R^2F/D_p, \\ \gamma &= N_{Fr}^{-2}N_{DF}^{-1}N_m\left(1-\frac{\bar{\rho}}{\bar{\rho}_p}\right)\cos\theta = (2R^2g\cos\theta/D_pu_0)\left(1-\frac{\bar{\rho}}{\bar{\rho}_p}\right), \\ \eta &= N_{Fr}^{-2}N_{DF}^{-1}N_m^2\left(1-\frac{\bar{\rho}}{\bar{\rho}_p}\right)\sin\theta = (2Rg\sin\theta/D_pF)\left(1-\frac{\bar{\rho}}{\bar{\rho}_p}\right), \end{aligned}\right\} \tag{8.21}$$

and the dimensionless numbers are:

The electrodiffusion numbers correlating transport by electrostatic repulsion to that by diffusion

$$N_{ev} = \left(\frac{\rho_{p0}}{4\varepsilon_0}\right)(q/m)^2\left(\frac{R^2}{D_p}\right)^2.$$

The diffusion response number, correlating relaxation time and diffusion time

$$N_{DF} = [(D_p/F)/R^2].$$

The Froude number, correlating inertia force to gravity force,

$$N_{Fr} = \frac{u_0}{\sqrt{2Rg}}$$

and the momentum transfer number, correlating relaxation time to convection time

$$N_m = u_0/RF.$$

The motion of the fluid phase is governed by the Reynolds number, N_{Re}

$$N_{Re} = u_0 R/\bar{\nu}.$$

The boundary conditions are now

$$\left.\begin{aligned} r^* &= 1-(a/R), \quad \text{or} \quad y^* = a/R, \\ V^* &= V^*(1, \varphi), \end{aligned}\right\} \tag{8.22}$$

and

$$u^*_{pw} = -N_{Kp}(\partial u^*_p/\partial r^*)|_{r*=1-(a/R)}. \tag{8.23}$$

For a conducting pipe, V^*_R is as specified before, and the Knudsen number for particle–fluid interaction is

$$N_{Kp} = L_p/R. \tag{8.24}$$

Therefore the state of the particulate phase in fully developed pipe flow with simultaneous electrostatic and gravity effects is defined by N_{Re}, N_{ev}, N_{DF}, N_{Fr}, N_m, and N_{Kp}. Equations (8.18) to (8.20) are applicable to pipe flow at any inclination to the direction of gravity. We have the following situations:

Case 1. For uncharged particles $V^* = 0$, and negligible gravity effect $\eta = 0$, $\rho^* = 1$, velocity distribution is given by eqn. (8.18) with $\partial/\partial\varphi = 0$.

Case 2. $V^* = 0$, but substantial η, non-depositing gravity flow is governed by eqns. (8.18) and (8.19). Vertical upward or downward flow is governed by the sign of γ with $\eta = 0$, $\rho^* = 1$ but $u^*_p = u^*_p(r^*, \varphi)$. For horizontal flow, $\rho^* = \rho^*(r^*, \varphi)$ but still we have $u^*_p = u^*_p(r^*, \varphi)$ even though $\gamma = 0$, because of dependence on ρ^*.

Case 3. For finite particle charge but negligible η, we have the case of non-depositing flow of a charged suspension without gravity effect, $\partial/\partial\varphi = 0$ in eqns. (8.18) to (8.20).

Case 4. When both gravity and electrostatic charge effects are present, complete solution is needed. With case 2 as a simplification, case 4 still has symmetry with respect to the vertical plane through the pipe axis.

Case 3 has been solved rigorously. Substituting $x = r^2$, $f = 2\alpha \int \rho_p r dr$, eqns. (8.19) and (8.20) combine to (*'s are dropped for simplicity, but all quantities are dimensionless)

$$x(d^2f/dx^2) = f(df/dx), \tag{8.25}$$

with $x = 0$, $df/dx = \alpha, f = 0$. Equation (8.25) is integrated to give

$$\rho_p = df/dx = f' = \left(1 - \frac{\alpha x}{2}\right)^{-2} = \sum c_n x^n, \tag{8.26}$$

with $c_n = (n+1)2^{-n}\alpha^n$. The density at $r = R$ is given by (for $R \gg \alpha$)

$$\rho_{pw} = f'(\alpha) = \left(1 - \frac{\alpha}{2}\right)^{-2}, \tag{8.27}$$

and the gradient of density is $(d\rho_p/dr)_{r=1} = 2\alpha f''(\alpha)$, and the mean density $\langle\rho_p\rangle$ across the pipe is $\alpha^{-1}f(\alpha)$. Moreover, $\rho_p = 1$ for $\alpha = 0$, i.e. when charges are absent. Electrostatically sedimentary flow or deposition will occur when ρ_{pw} reaches the magnitude of dense solid packing.

The velocity distribution u_p is given by eqn. (8.18) in the form

$$(d/dx)[x\rho_p(du_p/dx)] - (\beta/4)\rho_p u_p = -(\beta/4)\rho_p u, \tag{8.28}$$

and u_p is given by its particular solution for u given by a polynomial to avoid the slow convergence of a binomial expansion

$$u = y^{1/7} \sim b_0+b_1x+b_2x^2+b_3x^3+b_4x^4 \tag{8.29}$$

where $b_0 = 1$, $b_1 = -0.505$, $b_2 = 0.635$, $b_3 = 0.539$, and $b_4 = -0.589$; with ρ_p given by eqn. (8.26) and the boundary conditions ($R \gg a$)

$$u_p(1) = -2N_{Kp}du_p/dx|_{x=1} = u_{pw}, \tag{8.30}$$

the condition $du_p/dr = 0$ is satisfied by expansion in x. u_p is given by the series expansion

$$u_p = a_0+a_1r^2+a_2r^4+a_3r^6+\dots \tag{8.31}$$

with the coefficients a_{n+1} determined by eqn. (8.28) giving

$$a_{n+1} = (\beta/4)(n+1)^{-2}[(a_n-b_n)c_0+(a_{n-1}-b_{n-1})c_1+\dots+(a_0-b_0)c_n]- \\ -(n+1)^{-1}[na_nc_1+(n-1)a_{n-1}c_2+\dots+(1)a_1c_n], \tag{8.32}$$

where $c_0 = 1$. This recurrence relation gives a series of rapid convergence even for large β [for $\beta = 40$, convergence is reached after terms of $(\beta/4)^{10}/(10!)^2$], a_0 is then determined from

$$u_{pw} = \sum_{n=0} a_n = -2N_{Kp}\sum_{n=0} na_n \tag{8.33}$$

giving

$$a_0 = a_0(\alpha, \beta, N_{Kp}) \tag{8.34}$$

and

$$u_{pw} = u_{pw}(\alpha, \beta, N_{Kp}). \tag{8.35}$$

The numerical results are illustrated in Fig. 10, showing that greater lag of core velocity ($u_0 = u_{p0}$) and greater slip velocity (u_{pw}) occurs at smaller β (=4) than at larger β (=40),

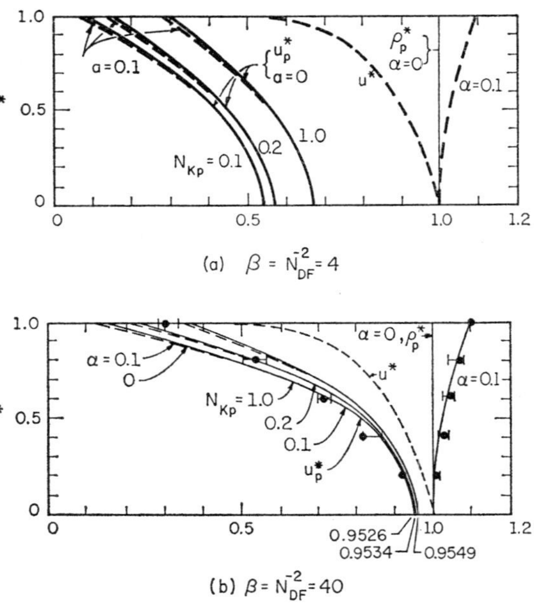

FIG. 10. Velocity and density profiles in fully developed turbulent pipe flow of a suspension (0 measured value for solid–gas mass ratio of 0.45, u_0 = 138 fps, 5 in. pipe; — range for other mass ratios). Measurements made by G. J. Trezek.

but for the same β, greater slip velocity u_{pw} occurs at larger N_{Kp} as expected. At $\beta = 0[10^2]$, $u_0 - u_{p0}$ is no longer noticeable, although it is still finite. The effect of large α on u_{pw} is similar to an increase in ρ_{p0}. The trends in Fig. 10 are confirmed by experimental results for horizontal pipe flow of suspensions of magnesia particles in air at room conditions.[111] These relations confirm the fact that while the fluid is in turbulent motion, the particles are in a state of laminar slip motion. Because of the lack of mutual interaction among the particles, the particle motion is uncorrelated.

For other cases outlined on page 117, numerical solution of eqns. (8.18) to (8.20) were obtained. It is seen that depending on the extent of the gravity effect the net motion of the particle phase may take the form of cocurrent, fluidized, or counter current motion with respect to the fluid phase.[1] Figure 11 illustrates a case of simultaneous electrostatic and gravity effects.

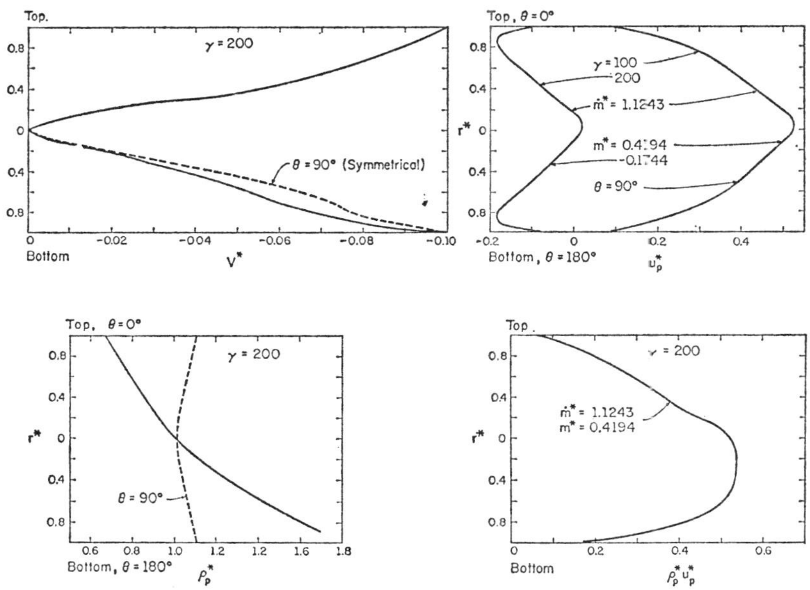

FIG. 11. Distributions in pipe flow with various gravity and charge effects. $\alpha = 0.2$, $\eta = 1.0$, $\beta = 100$, $N_{Kp} = 0.1$. Computations performed by S. K. Tung.

8.3. Measurement on charged suspensions

Experimental results[1,111] for glass and magnesia particles in air were in the range of $0.02 < \alpha < 0.2$ for q/m of the order of 10^{-4} coul/kg. A recent series of experiments[74] at 15 different mass flow rates of coal in the range of 0.008 to 0.60 gm/cm^2 sec in a 5 in. diameter copper pipe have been performed with mass flow obtained from an isokinetic sampling probe.[113] Maximum air velocity was 115 ft/sec. The overall mass flow ratio of particles to air $\dot{m}^*$ ranged from 0.0021 to 0.658 and, correspondingly, the mean inter-particle spacing varied from $120a$ to $20a$, where a is the mean particle radius. The mass flow distribution was symmetrical about the duct axis at three different air velocities at low mass flow ratio but showed sedimentation at high mass flow ratio. Both overall mass flow ratio and overall mass ratio of coal particles to air were computed for various core velocities. The relation is shown in Fig. 12. This relation characterizes the difference in mean velocities of the phases. The electrodiffusion numbers were determined from the measured electric charges carried by the particles; α was shown to vary from 0.06 to 0.30.

From the measured particle velocity profile, the Knudsen number for particle–fluid interaction N_{Kp} was computed from eqn. (8.33) or from the condition at the wall following eqn. (8.13). Their agreement is shown in Fig. 13. The magnitude of N_{DF} in the experiments ranged from 0.76×10^{-4} at low $\dot{m}^*$ to 1.70×10^{-4} at large $\dot{m}^*$.

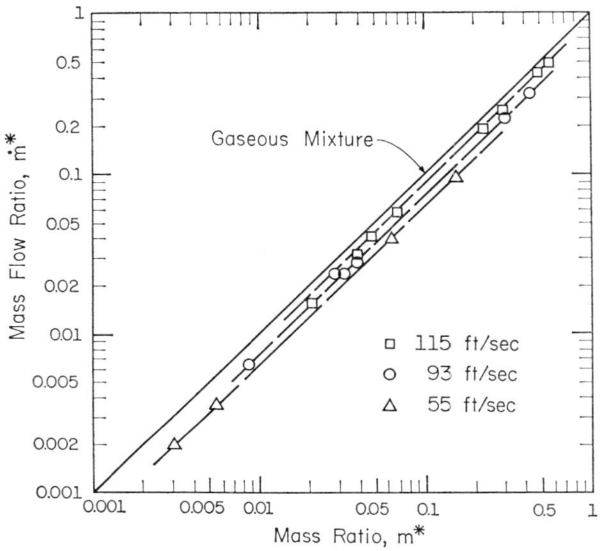

FIG. 12. Mass ratio versus mass flow ratio coal dust in air. Measurements made by L. Cheng and S. K. Tung.

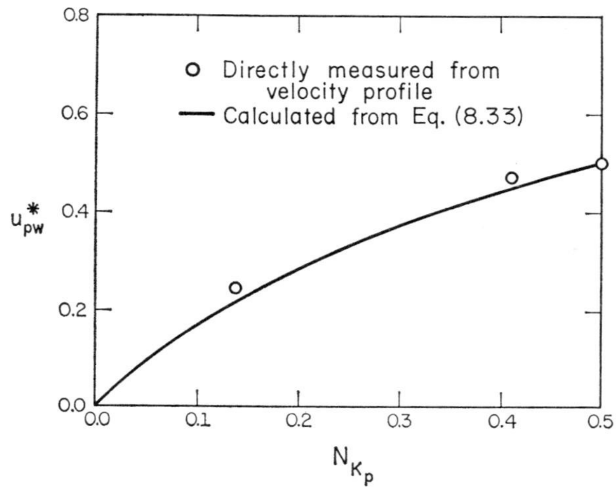

FIG. 13. The Knudsen number for particle–fluid interactions. Measurements made by L. Cheng.

8.4. Laminar flow of a liquid–solid suspension

Laminar flow of a gas–particle suspension is readily calculated as in section 8.2. However, the Magnus effect is expected to be significant. This is best illustrated by a liquid–solid suspension. Here, the effect of shear motion given by eqn. (8.6) is more significant than that of electric charges on particles produced by potential difference with the wall.(73) With $\boldsymbol{f}_L$ but for a suspension of neutral buoyancy ($\bar{\rho}_p = \bar{\rho}$), eqns. (8.2) and (8.4) take the form [all quantities are dimensionless as defined by eqn. (8.16)](73)

$$\beta(u-u_p) = -r^{-1}(d/dr)r(du_p/dr)-(du_p/dr)(d \ln \rho_p/dr), \tag{8.36}$$

$$(N_S/\beta)(u-u_p-\Delta u)(2r)^{1/2} = d \ln \rho_p/dr, \tag{8.37}$$

where N_S is denoted as the shear response number, correlating the Magnus force to the inertia force

$$N_S = (3/4\pi)(6.46)(\bar{\rho}/\bar{\rho}_p)(u_0 R/\bar{\nu})^{1/2}(R/a) \tag{8.38}$$

for laminar motion $\Delta u = -(du/dr)(a/R) = 2(a/R)r$. The simplest solution is for the case where β is very small, i.e. the effect of diffusion is negligible, thus $u_p \to u$ and eqn. (8.37) gives

$$\rho_p = \exp\{-[4\sqrt{(2/5)}](N_S/\beta)(a/R)r^{5/2}\} \tag{8.39}$$

and the concentrations are highest in the middle of the pipe conforming to the predicted trend of Einstein and Jeffery.(114, 115) Noting that

$$(N_S/\beta)(a/R) = (6.46/6\pi)(N_{Re})^{-1/2}(N_{Sc})^{-1}(a^2/R^2)[1+(\bar{\rho}/2\bar{\rho}_p)], \tag{8.40}$$

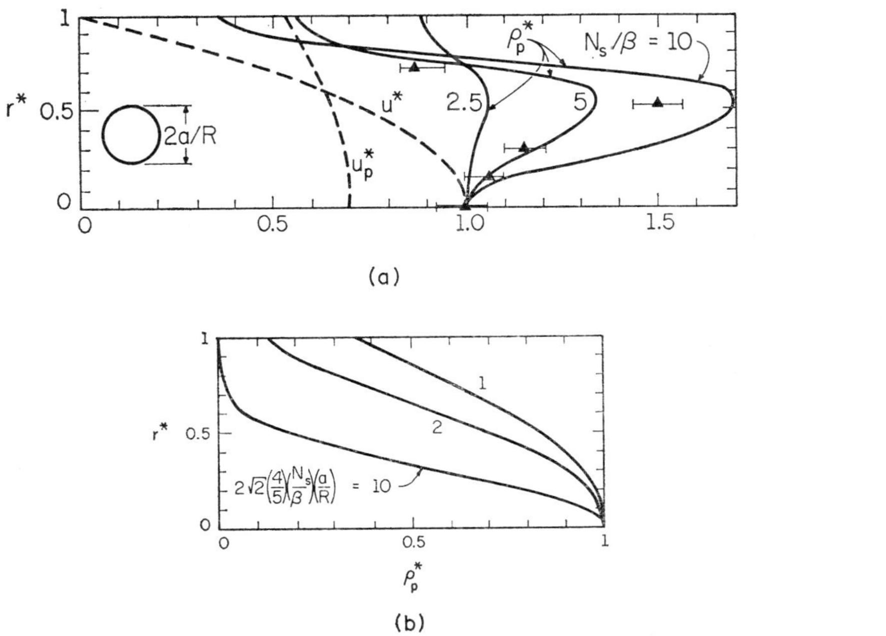

FIG. 14. Fully developed laminar pipe flow of a suspension. (a) Velocity and density profiles for $\beta(1-a_0) = 2/3$ in the range covered by the experiment of Segré *et al.* (Δ = experimental mean for Re = 7.3, $2a$ = 1.71 mm, u = 22.2 fps. (b) Density distribution in the range covered by the estimation of Einstein and Jeffery, $a/R \ll 1$.

the group $(N_S/\beta)(a/R)$ shows that a has to be sufficiently large (colloid range) for such a shear effect to arise; N_{Sc} is the Schmidt number of transport of particles:

$$N_{Sc} = \bar{v}/D_p. \tag{8.41}$$

The general situation, however, includes consistent approximation with both eqns. (8.36) and (8.37). Solution by series expansion gives, for $u_{p0}/u_0 = 0.7$, a maximum density ρ_p at $r = 0.6$ as experimentally determined by Segré and Silberberg.[116]

Figure 14 gives a comparison of the two ranges of $u_p = u$ and $u_{p0}/u_0 = 0.7$. Depending on the range of interaction the peak of particle density could occur at other radii. Also illustrated is the fact that depending on the value of (N_S/β), particles might appear to concentrate near the center of the pipe leaving a layer of clean fluid near the wall. Hence the results of Einstein and Jeffery and those of Segré and Silberberg apply to different ranges of motion given by various combinations of N_{Re}, N_{DF}, and N_S. Again, here the laminar state of motion of particles does not exclude their random motion.

CHAPTER 9

BOUNDARY LAYER MOTION

The boundary layer approximation to flow of a dilute charged suspension is applicable where fluid forces have a higher magnitude than electric forces,[1] the limit being an uncharged dilute suspension. Cases considered here are flow over flat plates and circular jets. Further, from earlier approximation,[1, 117] a rigorous solution of laminar flow of a suspension over a semi-infinite flat plate was given by Soo,[77] delineating the basic two-dimensional fluid–particle–boundary interactions.

9.1. Flow over a single flat plate[77]

The nature of non-equilibrium flow is seen from various quantities in the physical coordinates as shown in Fig. 15, obtained by extending the Blasius solution of laminar flow of a single phase fluid over a flat plate.[36] Using dimensionless quantities, we denote the coordinate

$$y^* = y/\sqrt{(\bar{v}/F)} \tag{9.1}$$

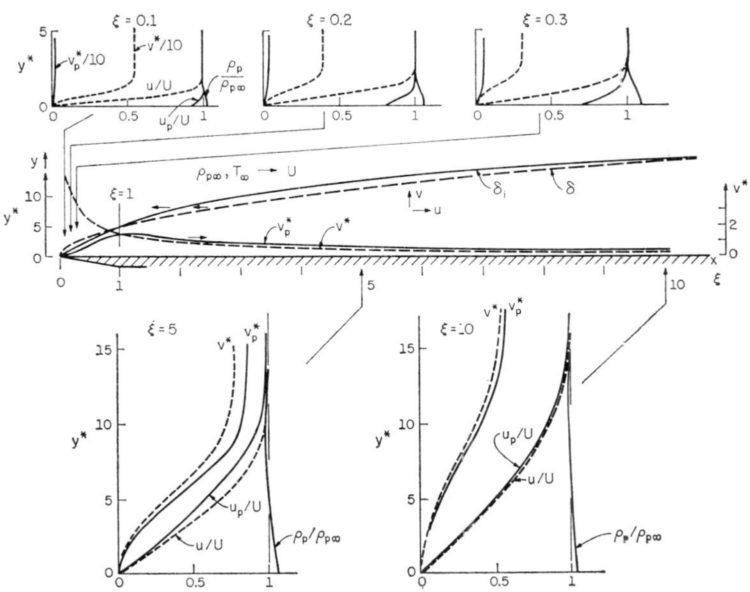

FIG. 15. Coordinates, boundary layer thicknesses, and distributions in laminar flow over a flat plate.

and, with $\xi = xF/U$, velocity of particles v_p in the y-direction as

$$v_p^* = \left[\frac{v_p}{U}\left(\frac{xU}{\bar{\nu}}\right)^{1/2}\right] \xi^{-1/2}, \tag{9.2}$$

and v^* is that for the fluid phase defined in a similar manner. In these relations and Fig. 15, U, $\rho_{p\infty}$, T_∞, $T_{p\infty}$ are the flow velocity of fluid, density of particles, fluid temperature, and particle temperature at $y = \infty$; u, v are components of the fluid velocity along the x, y direction; u_p, v_p for those of the particles; and $\bar{\nu}$ is the kinematic viscosity of the fluid. ρ, ρ_p are their densities; T, T_p are their temperatures; ρ is taken as a constant. The quantities v_p^*, v^* at $y = \infty$ at various ξ so determined are shown in Fig. 15 together with boundary layer thicknesses δ and δ_p given by y^* at $u/U = 0.99$. The transition at $\xi = 1$ is clearly seen. Figure 15 gives the distributions in ρ_p, u_p, u, v_p, v, δ_p, δ in a linear scale of ξ over the whole range of interest for $N_{Sc} = \bar{\nu}/D_p = 10$ and negligible L_p [eqn. (4.15)]. For the case of $G/F = 1$, $N_{Pr} = 1$, the profiles of reduced temperatures $(1-T_p^*)$ and $(1-T^*)$ are identical to those of u_p/U and u/U. It is seen that:

(1) Over the range $0 < \varepsilon < 1$, even for very small L_p, u_p is finite at the wall (slip) for large N_{Sc} and is slowed down by the fact that $u = 0$ at the wall and $N_{Kp} \sim 0$. Although the singularity in the Blasius solution gives $v^* \to \infty$ at $x = 0$, $v_p^* = 0$ initially for finite values of N_{Sc}, v_p^* is accelerated from zero in this range while v^* is decreasing until the point $\xi = 1$ is reached where $v_p^* = v^*$, $u_p = u$. $\rho_{pw}/\rho_{p\infty} > 1$ (ρ_{pw} being the particle density at the wall) prevails near the wall due to slowing down of the mixture. In this range $\delta_p < \delta$ as estimated before.[1]

(2) Equalization of velocities of components at $\xi = 1$ does not mean that the disparity of the components has died out. This is because v^* continues to decrease and in the range $\xi > 1$, $v_p^* > v^*$ due to the relaxation mechanism (inertia of particles p). $u_p < u$ at a given y^* for small ξ ($\xi < 1$) and the boundary layer thickness $\delta_p > \delta$ because $v_p^* > v^*$. For the same reason, $\rho_p/\rho_{p\infty}$ decreases below 1 in the boundary layer although $\rho_{pw}/\rho_{p\infty} > 1$ at the wall. It is seen that only at $\xi = \infty$, $u_p = u$, $v_p = v = 0$, $\rho_p = \rho_{p\infty}$, $\delta = \delta_p$. The decrease in ρ_{pw} shows that this range is diffusion-controlled.

(3) It is well known[36] that $(N_{Re})_x = xU/\bar{\nu} \sim 10^5$ for the transition from laminar to turbulent boundary layer. This relation gives transition at

$$\xi \gtrsim 10^5 \left(\frac{U^2}{\bar{\nu}F}\right)^{-1} = \xi_{tr} \tag{9.3}$$

for the laminar range. If we take, for example, $U = 100$ m/sec, $\bar{\nu} = 10^{-4}$ m^2/sec, $F = 10^5$ sec^{-1} (such as 1 μ aerosols in air). Therefore, in the former case, slip of p would not have considerably diminished and $\xi < 1$ before turbulence sets in at 10 cm from the leading edge, while in the latter case, the observable ranges include 1 mm for $\xi \leq 1$, and 10 cm for the range $1 < \xi < 100$.

(4) The previous items shows that, when transition to turbulence occurs within the range $0 < \xi < 1$, presence of component p in sufficient quantity would delay the transition since $v_p < v$. Yet, when transition to turbulence occurs beyond $\xi = 1$, component p enhances transition at lower $(N_{Re})_x$ than that in a classical fluid because $v_p > v$. Therefore, from the examples in item (3), transition to turbulence will be suppressed by 100 μ sand in air, while the 1 μ aerosols would aid the transition. This understanding serves to resolve the inconclusiveness of previous studies[118] on the effect of particles on the turbulence transition of the fluid phase.

These considerations are applicable to flow of large finite cloud of charged particles over a flat plate discussed by Soo.[1] In general

$$(N_{Re})_x \gg (N_{ev})_x$$

for negligible effect of electric charges; $(N_{ev})_x = (\rho_{p0}/4\pi\varepsilon_0)(q/m)^2(x^2/\bar{\nu})^2$. The magnitude of physical quantities can be seen in the case of $\bar{\nu} = 10^{-5}$ m^2/sec, $q/m = 10^{-4}$ coul/kg, $U = 100$ m/sec, electric field $E_y/_{y=0} = 10^6$ v/m at $x = 0.01$ m, giving a surface charge density of 10^6 ε_0. For an uncharged plate, the image charge would give rise to collection patterns such that the concentration is high on the downstream side. For a plate charged to the opposite sign, collection will tend toward the leading edge. The modification on boundary layer thickness will follow a similar trend. With the plate charged to the same sign, the boundary layer could exhibit a lower concentration at the wall.

The above effect is related to the influence of a charged cloud of particles on the aerodynamic behavior of, say, the wing of an aircraft, and the above consideration is applicable to the estimation of the change in lift and drag in its passage through a rainstorm. The electrical interaction of an aircraft with a charged cloud of rain is further expected to affect the flight safety. On the one hand, an uncharged large aircraft may receive the discharge of a thundercloud much as a ground object.[10] On the other hand, collection of charges due to electroaerodynamic interaction may occur in the same manner as a large droplet illustrated in section 7.3, and the adequacy of discharge points has to be assured. To be sure, lightning hazard of an aircraft is a very rare occurrence, and if it did occur, little clue would have been left. Tests with a F-100F aircraft showed that high voltage discharge of 190 kA punctured a canopy with an aluminium foil protection strip. Flight tests showed that nose boom, wingtips, and the tips of the stabilizers were often struck. Cloud-to-cloud, or in-cloud strokes, or both, were observed. No inducting mechanism was needed to encounter these strokes. Largest peak currents recorded were 12 kA. Cloud-to-ground strikes through the aircraft were not attained.[119] It is expected that the hazard of flight through a rainstorm increases with the size of the craft.

9.2. Turbulent flow into a channel

The case of an infinite homogeneous charged suspension in a system represented by Fig. 15 is physically unattainable and mathematically divergent. For a definitive study of turbulent boundary layer motion of a charged suspension, Stukel and Soo[120] chose the system of flow into a channel formed by two parallel flat plates as shown in Fig. 16.

With the two parallel flat plates having sharp leading edges situated at a distance $2b$ apart, the coordinate x is aligned with the flat plates and the coordinate y normal to it. The corresponding components of velocities are u, v for the fluid and u_p and v_p for the particles. Each of the particles has radius a with a material density $\bar{\rho}_p$, and is suspended in a fluid having a material density of $\bar{\rho}$. The suspension is dilute such that the density of the particulate suspension $\rho_p \ll \bar{\rho}_p$; hence $\rho \sim \bar{\rho}$.

For the system defined above, with flow in the subsonic range, the carrier fluid may be treated as incompressible. However, the dependence of ρ_p on the transport processes must be taken into account. We therefore have the following continuity relation:

$$(\partial u/\partial x)+(\partial v/\partial y) = 0, \tag{9.4}$$

$$(\partial/\partial x)(\rho_p u_p)+(\partial/\partial y)(\rho_p v_p) = 0. \tag{9.5}$$

The equations for the x-components of momenta of the phases with boundary layer simplifications are:

$$\rho u(\partial u/\partial x)+\rho v(\partial u/\partial y) = -(\partial P/\partial x)-KF\rho_p(u-u_p)+(\partial\tau/\partial y), \tag{9.6}$$

$$u_p(\partial u_p/\partial x)+v_p(\partial u_p/\partial y) = F(u-u_p)+E_x(q/m)+\rho_p^{-1}(\partial\tau_p/\partial y), \tag{9.7}$$

where τ, τ_p are the shear stresses, P is the static pressure, and E_x is the electric field in the x-direction. Equation (9.6), through an effectiveness parameter K,[1] accounts for a

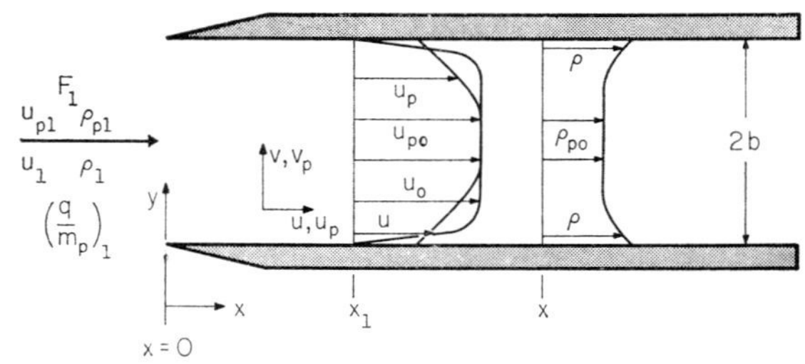

FIG. 16. System for studying turbulent flow into a channel and coordinates.[120]

momentum transfer from the freely suspended particles to the fluid. The y-component of particle motion is given by

$$u_p(\partial v_p/\partial x)+v_p(\partial v_p/\partial y) = F(v-v_p)+E_y(q/m). \tag{9.8}$$

The E field is given by the Poisson equation

$$(\partial E_x/\partial x)+(\partial E_y/\partial y) = \rho_p(q/m)/\varepsilon_0, \tag{9.9}$$

where ε_0 is the permittivity.

Because of symmetry, we can outline the boundary conditions at one of the plates, such that, at $x = 0$, $u = u_p = u_1$, $v = v_p = 0$, $\rho_p = \rho_{p1}$; and at $x > 0$: $y = b$, $u = u_0(x)$, $u_p = u_{p0}(x)$, $v = v_p = 0$, $\rho_p = \rho_{p0}(x)$, $E_y = 0$. At $y = 0$: $u = 0$, $u_p = u_{pw}(x)$, $v = 0$, $\rho_p = \rho_{pw}(x)$. Equation (9.7) and the slip condition further give[77]

$$u_{pw} = u_1[1-\xi]+L_p(\partial u_p/\partial y)_{y=0}, \tag{9.10}$$

where the first term on the right-hand side accounts for slip at the inlet, with $\xi = Fx/u_1$ at $0 < x < u_1/F$, and $\xi = 1$ at $x > u_1/F$, and the second term account for finite interaction length defined before. Further, inside the channel, it is assumed $(\partial E_y/\partial y) \gg (\partial E_x/\partial x)$. Hence for given τ for the turbulent fluid and τ_p for a dilute suspension[112] given by

$$\tau_p \sim \rho_p D_p(\partial u_p/\partial y). \tag{9.11}$$

Equations (9.4) to (9.9) are solved for u, v, u_p, v_p, ρ_p, and E for given ρ, ρ_{p1}, u_1, subject to the above boundary condition and the stated assumptions.

Furthermore, for dilute suspension, eqn. (9.5) can be replaced by the equation of diffusion under field forces in the form

$$u\frac{\partial\rho_p}{\partial x}+v\frac{\partial\rho_p}{\partial y} = -\frac{\partial}{\partial y}\left[\left(\frac{q}{m}\right)\frac{E_y\rho_p}{F}\right]+\frac{\partial}{\partial y}\left(D_p\frac{\partial\rho_p}{\partial y}\right) \tag{9.12}$$

according to eqns. (6.12) to (6.15) and $|E_y| \gg |E_x|$ in eqn. (9.9).

Reduction of the above equations to those of momentum integrals was carried out by introducing, from eqns. (9.4) and (9.5),

$$v = -\int (\partial u/\partial x)\,dy, \tag{9.13}$$

$$\rho_p v_p = -\int (\partial \rho_p u_p/\partial x)\,dy + \rho_{pw} v_{pw}, \tag{9.14}$$

where v_{pw} accounts for deposition of particles (or erosion from the wall). Further, the following momentum integrals for boundary layer thickness δ and δ_p of the fluid and particle phase were introduced:[1, 120]

$$\delta^* = \int_0^{\delta} \left(1 - \frac{u}{u_0}\right) dy, \qquad \theta = \int_0^{\delta} \frac{u}{u_0}\left(1 - \frac{u}{u_0}\right) dy,$$

$$\delta_p^* = \int_0^{\delta_p} \left(1 - \frac{\rho_p u_p}{\rho_{p0} u_{p0}}\right) dy, \qquad \theta_p = \int_0^{\delta_p} \frac{\rho_p u_p}{\rho_{p0} u_{p0}}\left(1 - \frac{u_p}{u_{p0}}\right) dy,$$

and a gas–particle interaction thickness to correlate particle–fluid interactions,

$$\delta_{pg} = \int_0^{\delta} \left(1 - \frac{\rho_p u}{\rho_{p0} u_0}\right) dy.$$

Equations (9.6), (9.7), and (9.12) were reduced to

$$(d/dx)(u_0^2\theta) + \delta^* u_0 (du_0/dx) + KF(\rho_{p0}/\rho)(u_0\delta_{pg} - u_{p0}\delta_p^*) = \frac{\tau}{\rho}, \tag{9.15}$$

$$(d/dx)(\rho_{p0} u_{p0}^2 \theta_p) + \rho_{p0}\delta_p^* u_{p0}(du_{p0}/dx) + \rho_{pw} v_{pw}(u_{p0} - u_{pw}) = F\rho_{p0}(u_0\delta_{pg} - u_{p0}\delta_p^*) + $$
$$+ \rho_{pw} D_{pw}(\partial u_p/\partial y)_w, \tag{9.16}$$

$$u_0(\delta_{pg} - \delta_p)(d\rho_{p0}/dx) + \rho_{p0}(d/dx)[u_0(\delta_{pg} - \delta^*)] = \left(\frac{\rho_{pw} - \rho_{p0}}{\varepsilon_0 F}\right)\left(\frac{q}{m}\right)^2 \rho_{p0}(b - \delta_p) +$$
$$+ \left(\frac{q}{m}\right)^2 (\rho_{pw}/F\varepsilon_0) \int_0^{\delta_p} \rho_p\,dy + D_{pw}(\partial \rho_p/\partial y)_w, \tag{9.17}$$

for the transfer of momentum of the phases and the distribution of particle density.

As in the case of single phase turbulent boundary layers, evaluation of the integrals has to be made semi-empirically. Experiments with a suspension of 10 μ magnesia in air flowing over the inlet of a channel formed by two flat plates was made for various flow velocities, plate gap widths, and mass flow ratios of solids to air. Experiments were carried out in a 12 in. by 12 in. section wind tunnel with flow velocities up to 120 fps, plate gap widths of 1/4, 1, and 2 in., and mass flow ratios up to 0.1 lb particles per lb air. Primary measurements included the determination of the particle and air velocities, the particulate mass flow and density distributions, and the particle size distribution as influenced by the flow response.[113]

Measurements showed that the motion of the gaseous phase satisfies the turbulent boundary layer theory.[36] The displacement thickness was shown to follow for 1/7 the fluid velocity profile,

$$\frac{\delta^*}{b} = 0.0463\varepsilon\left(\frac{u_1}{u_0}\right)^{1/5}, \tag{9.18}$$

where $\varepsilon = (x/b)^{4/5}(u_1 b/\bar{v})^{-1/5}$ and

$$\frac{u_0}{u_1} = 1 + K_1\varepsilon + K_2\varepsilon^2 + \ldots, \tag{9.19}$$

where $K_1 = 0.0463$, $K_2 = 0.001712, \ldots$, for the core velocity. Further, the momentum thickness is given by

$$\theta/b = 0.036\varepsilon\left(\frac{u_1}{u_0}\right)^{1/5}. \tag{9.20}$$

The core velocity of the particles is readily shown by integrating eqn. (9.7) in the fluid core to give, with eqn. (9.19),

$$\frac{u_{p0}}{u_{p1}} = 1 + \left(\frac{5Fx}{u_1}\right)\left(\frac{K_1}{9}\varepsilon + \frac{K_2\varepsilon^2}{13} + K_3\frac{\varepsilon^3}{17}\right) - \left(\frac{5Fx}{u_1}\right)^2\left(\frac{K_1}{126}\varepsilon + \frac{K_2}{234}\varepsilon^2 + \frac{K_1^2}{5\cdot 243}\varepsilon^2 \ldots\right), \quad (9.21)$$

F determined from this relation utilizing the results ranging from 550 to 700 sec^{-1}, for particle sizes varying from 13.5 to 11.5 μ. With the reduced coordinate x^*,

$$x^* = (Fb/u_1)(x/b), \quad (9.22)$$

the dependence on the relaxation parameter F in particle motion was noted. It was found that within the accuracy of measurement L_p is simply a function of u_1 for the present system.

The particle velocity profile is satisfied by the relation

$$\frac{u_p - u_{pw}}{u_{p0} - u_{pw}} = \frac{\gamma}{\gamma-1}\left(\frac{y}{\delta_p}\right) - \frac{1}{\gamma-1}\left(\frac{y}{\delta_p}\right)^\gamma. \quad (9.23)$$

Over the range of the present investigation, namely $\dot{m}_1^*$ varying between 0.01 to 0.125, with u_{p1} varying from 67 to 100 fps, γ was experimentally found to be 2. Hence, the above equation becomes

$$\frac{u_p - u_{pw}}{u_{p0} - u_{pw}} = 2\left(\frac{y}{\delta_p}\right) - \left(\frac{y}{\delta_p}\right)^2. \quad (9.24)$$

Equation (9.24) is shown in Fig. 17 along with the experimentally determined profiles. The agreement between the power law profile and the data is very good over the range of the experiments.

If γ is set equal to 2, one obtains from eqn. (9.10), neglecting $(1-\xi)$,

$$\frac{u_{pw}}{u_{p0}} = 1\bigg/\left(\tfrac{1}{2}\frac{\delta_p}{L_p} + 1\right). \quad (9.25)$$

Figure 18 gives a comparison of u_{pw}/u_{p0} calculated from eqn. (9.25) with experimentally determined values of δ_p, L_p, and u_{pw}/u_{p0}.

The particle density distribution of the particle phase was shown to follow from eqn. (9.19) and eqn. (9.12) in the core ($u = u_0$, $v = 0$).

$$\frac{\rho_{p1}}{\rho_{p0}} = 1 + \left(\frac{q}{m}\right)^2\left(\frac{b\rho_{p1}}{u_1 F\varepsilon_0}\right)\left(\frac{x}{b}\right)[1 - 0.0257\varepsilon + 0.0001649\varepsilon^2 - \ldots]. \quad (9.26)$$

The charge to mass ratio was at the level of 10^{-3} coul/kg. Similar to the particle velocity, a core region is seen to exist for the concentration distribution.

The wall density increases with the mass flow ratio. A mass balance at the wall gives

$$(\rho_p v_p)_x = -\frac{d}{dx}\int_0^b \rho_p u_p(x, y)\,dy, \quad (9.27)$$

the deposition flux at the wall (Fig. 19). The trend of an earlier computation was verified.[1,93]

Particle size distributions were determined from isokinetically collected samples on glycerine-treated glass slides. Micrographs were obtained for two conditions. Very close to the inlet, the average particle size at the wall was found to be about 15% smaller than at the center. At 2–3 in. the average sizes became equal, while at 12 in. the average particle size at the wall was 40% less than at the center position.

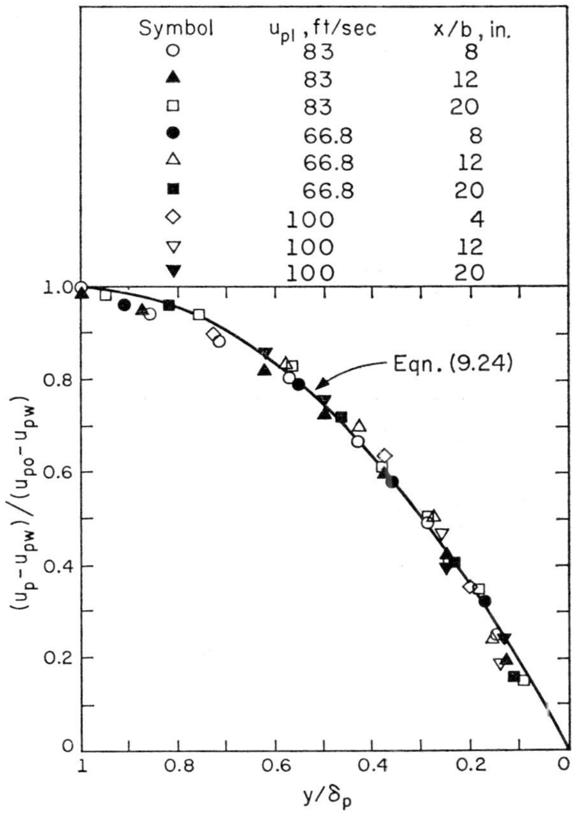

FIG. 17. Particle velocity distribution.[120] Measurements made by J. J. Stukel.

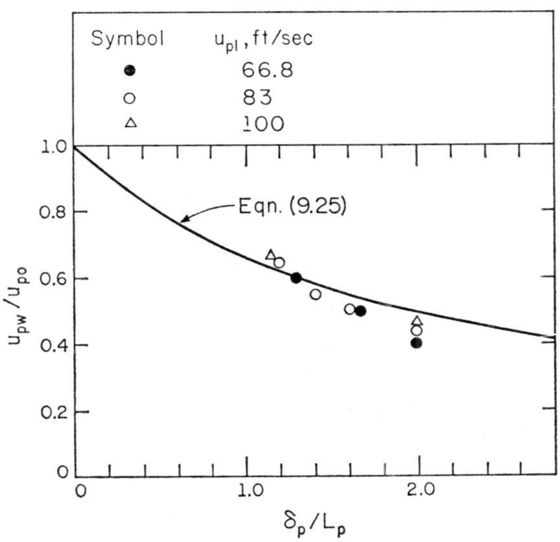

FIG. 18. Particle slip velocity as a function of δ_p/L_p.[120] Measurements made by J. J. Stukel.

Figures 20, 21, and 22 give the variation of δ_p^*, θ_p, and δ_{pg}. It is known in the region $0 < x < (u_1/F)$ that the particulate displacement thickness varies as

$$\delta_p^* \sim \left[\left(\frac{\bar{v}}{u_1 b}\right)\left(\frac{x}{b}\right)\right]^{1/2}\left(\frac{Fb}{u_1}\right)\left(\frac{x}{b}\right).$$

Now at $x = (u_1/F)$ the above becomes

$$\frac{\delta_p^*}{b} \sim \sqrt{\left(\frac{\bar{v}}{Fb^2}\right)}.$$

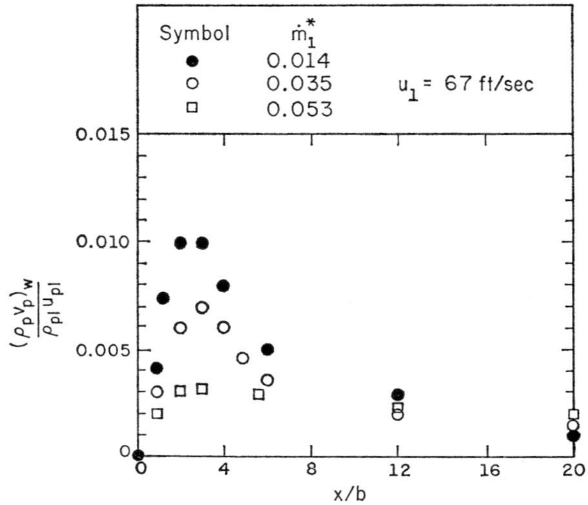

FIG. 19. Wall deposition flux for varying $\dot{m}_1^*$ and $u_1 = 67$ fps. Measurements made by J. J. Stukel.

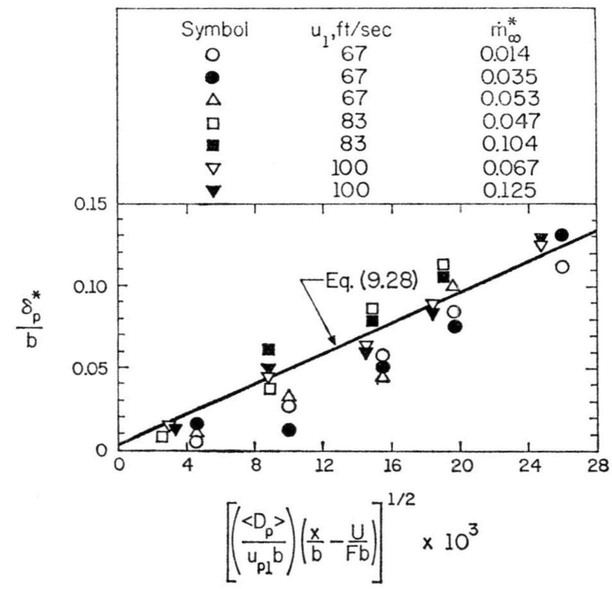

FIG. 20. Variation of δ^*.[120] From measurements made by J. J. Stukel.

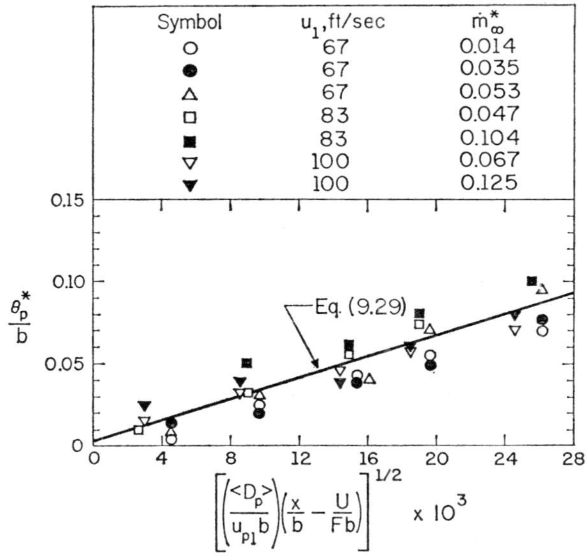

FIG. 21. Variation of θ_p.(120) From measurements made by J. J. Stukel.

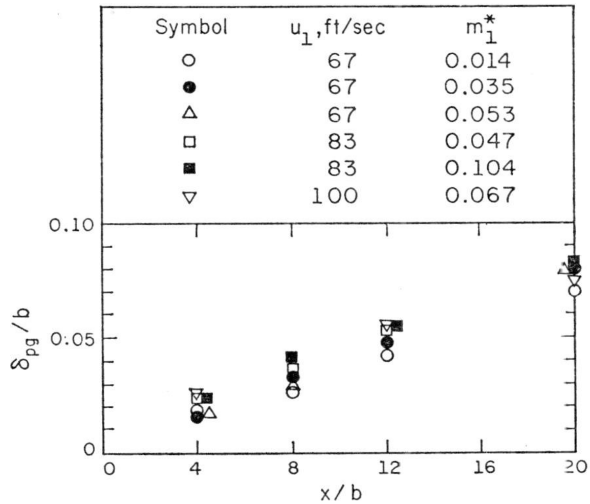

FIG. 22. Variation of δ_{pg}.(120) From measurements made by J. J. Stukel.

For the range of the present investigation, it is expected that δ_p^*/b and θ_p/b evaluated at $x = u_1/F$ to have magnitudes ranging from 0.002 to 0.008 assuming the constant of proportionality to be of the order of 1.

Since this is the case, finite values for the thickness δ_p^* and θ_p are anticipated at $x = u_1/F$ ($\varepsilon_p = 0$) and are noted in Figs. 20–22. In the region $x \geq u_1/F$, δ_p^* and θ_p are seen to correlate in a manner similar to laminar flow, namely,

$$\frac{\delta_p^*}{b} = 5.25\varepsilon_p + 0.0044, \tag{9.28}$$

$$\frac{\delta_p}{b} = 3.19\varepsilon_p + 0.0028, \tag{9.29}$$

where

$$\varepsilon_p = \left[\left(\frac{\langle D_p \rangle}{u_{p1} b}\right)\left(\frac{x}{b} - \frac{u_1}{Fb}\right)\right]^{1/2}.$$

Further, δ_p^*/θ_p is approximately constant as in laminary boundary layer flow. This again confirms the trend of the particle phase to have properties of laminar slip motion. δ_{pg}, however, correlating particle motion to fluid motion, does not show this trend.

Scaling relations for particulate flow can be found by rewriting eqns. (9.16) and (9.17) in the form

$$\begin{aligned}(d/dx^*)[\rho_{p0}^* u_{p0}^* u_{p0}^*(\theta_p/b)] + (\delta_p^*/b)\rho_{p0}^* u_{p0}^*(du_{p0}^*/dx^*) + \rho_{pw}^* v_{pw}^*(u_{p0}^* - u_{pw}^*) \\ = (Fb/u_1)[(u_0^*\delta_{pg} - u_{p0}^*\delta_p^*)/b] + \rho_{pw}^*(D_{pw}/u_1 b)(\partial u_p^*/\partial y^*)_w,\end{aligned} \tag{9.30}$$

$$\begin{aligned}u_0^*[(\delta_{pg} - \delta_p)/b](d\rho_{p0}^*/dx^*) + \rho_{p0}^*(d/dx^*)[u_0^*(\delta_{pg} - \delta^*)/b] \\ = (q/m)^2(\rho_{p1} b/u_1 F\varepsilon_0)\rho_{pw}^* \int_0^{\delta_p/b} \rho_p^*\, dy^* + (D_{pw}/u_1 b)(\partial\rho_p^*/\partial y^*)_w + \\ + \left(\frac{q}{m}\right)^2\left(\frac{\rho_{p1} b}{u_1 F\varepsilon_0}\right)\rho_{p0}^*(\rho_{pw}^* - \rho_{p0}^*)\left(1 - \frac{\delta_p}{b}\right),\end{aligned} \tag{9.31}$$

where $u^* = u/u_1^* = u_p^* = u_p/u_1$, $\rho_p^* = \rho_p/\rho_{p1}$, $y^* = y/b$, $x^* = x/b$.

As can be seen from the above relations, the parameter $N_{Re} = u_1 b/\nu$ normally used for scaling is not adequate for suspension flows. In addition, the above equations give, as in section 8.2,

$$N_m \equiv \frac{u}{bF},\quad N_{ev} \equiv (q/m)^2 \frac{\rho_{p1} b^4}{\varepsilon_0 D_p^2}$$

and diffusion-convection number

$$N_D \equiv \frac{D_{pw}}{u_1 b}$$

correlating the diffusion time to the convection time. Further, for the present system,

$$\left(\frac{q}{m}\right)^2\left(\frac{\rho_{p1} b}{u_1 F\varepsilon_0}\right) = N_{ev} N_m^{-1} N_D^2$$

is a suitable lumped correlation parameter.

9.3. Electrohydrodynamic jets

Distributions of velocity and density in a neutral particulate suspension or a disparate molecular mass mixture in both laminar and turbulent axisymmetric jets are known.[1, 117] Soo[121] treated hydrodynamic jets, including the space charge effect existing in a jet of charged particulate suspensions. This recent study concerns distribution in a jet with a charged trace component, the principal fluid component is neutral and incompressible, and the background fluid is identical to the principal fluid of the jet. The principal fluid jet is as defined by Schlichting and Tollmien for the laminar and turbulent ranges.[36] External field is absent. Since the relaxation controlled range of even a laminar jet occurs over a short distance from an orifice,[1] only the diffusion controlled range was considered. The forces acting on the trace component (ions or charged aerosols) are such that:

Fluid (drag) force > electrostatic force > inertia force.

For a dilute suspension, the velocity of the principal fluid is unaffected by the presence of the trace components. Schlichting and Tollmien gave the following axial (x-direction) and radial (r-direction) components (u, v) of velocity distribution:[36]

$$u = Df/x\eta, \quad v = D(\eta f' - f)/x\eta, \tag{9.32}$$

where D denotes kinematic viscosity ν of the fluid in laminar motion, and apparent kinematic viscosity, or eddy diffusivity in turbulent flow, where $D \sim r_1 u_1/39$ as given by Tollmien and others.[36] In eqn. (9.32) r_1 is the orifice radius and u_1 is the uniform velocity at the exit of the orifice; other quantities are $\eta = r/x$, $\xi = \gamma\eta$, $\gamma = (3^{\frac{1}{2}}/4)(r_1 u_1/D)$, $f = \xi^2/X$, $f' = 2\gamma^2\eta/X^2$, and $X = 1 + (\xi^2/4)$. The above velocity distribution is valid for $x > (3^{\frac{1}{2}}/2)\gamma r_1$. The jet boundary r_s at a given x is given by[1]

$$(2/\sqrt{3})\gamma r_1/x = \xi_s^2/X_s$$

or

$$X_s^{-1} = 1 - [\gamma r_1/2\sqrt{(3)}x]. \tag{9.33}$$

Under the specifications of the present system and the boundary layer assumptions, diffusion and self-field effects are small in the x-direction when compared to those in the r-direction. The flux of the trace component consists of that due to diffusion and that due to electric self-field J_{pE}. For densities of principal and trace components ρ, ρ_p, in the suspension, for $\rho \gg \rho_p$, the diffusion equation takes the form

$$u\,\frac{\partial \rho_p}{\partial x} + v\,\frac{\partial \rho_p}{\partial r} = \frac{1}{r}\,\frac{\partial}{\partial r}\,r\left[D_p\,\frac{\partial \rho_p}{\partial r} - J_{pE}\right], \tag{9.34}$$

where D_p is the diffusivity of component p in the mixture. When the inertial force of the particle is small, the radial momentum equation gives

$$\rho_p u_p\,\frac{\partial v_p}{\partial x} + \rho_p v_p\,\frac{\partial v_p}{\partial r} = 0 = \rho_p F(v - v_p) + E_r \rho_p (q/m) = -FJ_{pE} + E_r \rho_p (q/m), \tag{9.35}$$

where q is the charge and m is the mass of each particle of the trace component and E_r is the electric field; F^{-1} is the relaxation time of momentum transfer from the fluid to the particle phase.[1,77,94]

E_r ($\gg E_x$) is given by the Poisson equation

$$r^{-1}(\partial/\partial r)(rE_r) = \rho_p (q/m)/\varepsilon_0, \tag{9.36}$$

where ε_0 is the permittivity (MKS system). Combining eqns. (9.34), (9.35), and (9.36), we get, for D_p = constant,

$$\frac{1}{D_p}\left(u\,\frac{\partial \rho_p}{\partial x} + v\,\frac{\partial \rho_p}{\rho r}\right) = \frac{1}{r}\,\frac{\partial}{\partial r}\,r\,\frac{\partial \rho_p}{\partial r} - \frac{1}{D_p \varepsilon_0 F}(q/m)^2\,\frac{1}{r}\,\frac{\partial}{\partial r}\left(\rho_p \int_0^r r\rho_p\,dr\right). \tag{9.37}$$

Equation (9.37) is solved by changing the independent variables from x, r to x, η with $(\partial/\partial x)_r = (\partial/\partial x)_\eta - (\eta/x)(\partial/\partial\eta)_x$, and $(\partial/\partial r)_x = (1/x)(\partial/\partial\eta)_x$, u, v, given by eqn. (9.32) and ρ_p given by

$$\rho_p = (\beta/x)[h_0(\eta) + x^* h_1(\eta) + x^{*2} h_2(\eta) + \ldots], \tag{9.38}$$

where the perturbation parameter $x^* = [(q/m)^2/F\gamma^2 D_p \varepsilon_0]\beta x$, β being a constant determined from the conservation of total flow of particles. For $D/D_p = N_{Sc} \equiv N$, the Schmidt number, different orders of x^* are given for boundary conditions $h'(0) = 0$, $h(\infty) = 0$, solution of equations of various orders of h_n give for $N_{Sc} > 1/2$:

$$\left.\begin{aligned} h_0 &= X^{-2N}, \\ h_1 &= (1-2N)^{-1}X^{-4N+1}, \\ h_2 &= -(1-2N)^{-2}X^{-2}\left[\int \frac{dX}{(X-1)^{6N-4}} - \int \frac{dX}{X-1}\right], \end{aligned}\right\} \tag{9.39}$$

which has integral values for $6N-4 = \alpha$, $\alpha \geq 2$,

$$h_2 = -(1-2N)^2 \left[\sum_{(\lambda-1)}^{(\alpha-1)} \frac{1}{\lambda X^{\lambda+2}} - \frac{\ln X}{X^2}\right], \tag{9.40}$$

and

$$\beta = (\rho_{p1} u_1 r_1^2/8D)(1+2N)(1-X_{ps0}^{-2N-1})^{-1}. \tag{9.41}$$

The perturbation parameter x^* is now given by

$$\begin{aligned} x^* &= (1-X_s^{-2N-1})^{-1}\left[\frac{\rho_{p1}}{\varepsilon_0}(q/m)^2 \frac{r_1^4}{D_p^2}\right]\left[\frac{D_p}{F r_1^2}\right]\frac{4}{3}\left(\frac{r_1 u_1}{\nu}\right)^{-1}\left(\frac{D}{\nu}\right)(1+2N_{Sc})\left(\frac{x}{r_1}\right) \\ &= (1-X_s^{-2N-1})^{-1} N_{ev} N_{DF} N_{Re}^{-1}\left(\frac{D}{\nu}\right)(1+2N_{Sc})\frac{x}{r_1}, \end{aligned} \tag{9.42}$$

where N_{ev}, the "electrodiffusion number", correlates electrostatic transport to fluid transport of particles. N_{DF}, the diffusion response number, correlates relaxation time to diffusion time of particles. N_{Re}, the Reynolds number, correlates inertia force to viscous force of the fluid. $D/\nu = 1$ for laminar motion, and $D/\nu = N_{Re}/39$ for a turbulent jet. In the case of a turbulent jet, x^* is unaffected by N_{Re}.

The spreading of a jet of a particulate suspension is illustrated in Fig. 23, comparing the cases with and without electric charges on particles. It is seen that $x^* = 0(10^{-1})$ is obtained at $x = 20$ cm. For $r_1 = 1$ cm, $\rho_p = 0.1$ kg/m^3, $(q/m) = 10^{-3}$ coul/kg, $D_p = 10^{-3}$ m^2/sec, $F^{-1} = 10^{-4}$ sec (1 μ aerosol in air at STP), in a turbulent jet. In the same study, ambipolar diffusion was also treated. Experimental verification of the above correlation was given by Lee and Soo.[122]

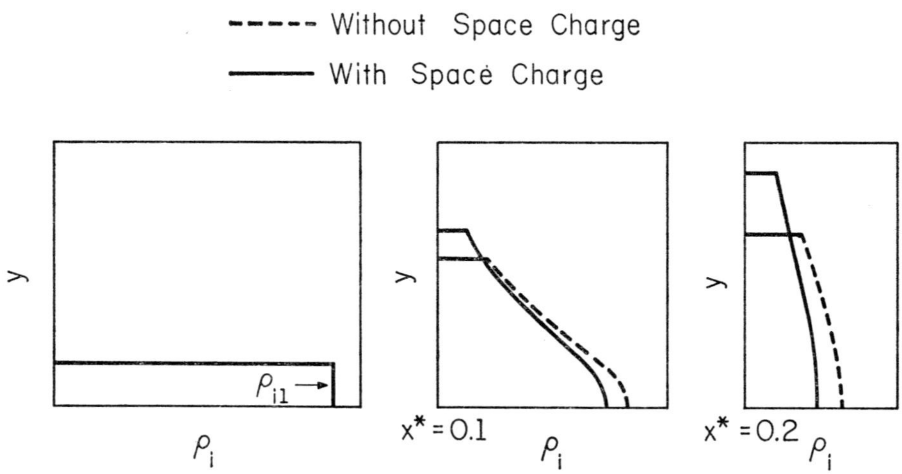

FIG. 23. Comparison of density distribution of particles with and without space charge effect ($N_{Sc} = 1$).

9.4. Dispersion in the atmosphere

Prediction of the effect of atmospheric dispersion of stack effluents is important to the control of environmental quality. The distribution of concentration of gaseous contaminants in a stack plume is given by the Sutton diffusion equation.[130] The rate of deposition or fall-out of particulates by gravity but neglecting diffusion was given by Bosanquet, Carey, and Halton.[130] A general treatment is desirable since the effects of gravity and uniform electric field (section 7.4) also influence rate of fall-out of heavy molecules and ions while diffusion also affects transport of small particulate matter. A general equation including both diffusion and fall-out was derived.[110] From this equation, ground-level flux of deposition was computed and charts were presented. Applicable cases are plumes of smoke stacks and exhaust of aircraft, and transient injection of pollutants (section 7.4).

For a smoke stack of exist radius r_1, at velocity w_1 into a cross wind of velocity u, and density ρ_{k1} of contaminant species k, the density distribution in the plume of species k, ρ_k, is given by solving the diffusion equation with field forces. It takes the form

$$\rho_k = \rho_{k1}\left(\frac{r_1}{h}\right)^2\left(\frac{w_1}{u}\right)\left(\frac{D_{ky}}{D_{kz}}\right)^{1/2}\left(\frac{1}{4x_k^*}\right)\exp\left[-\frac{1}{4x_k^*}\frac{y^2}{h^2}\frac{D_{kz}}{D_{ky}}\right]$$
$$\left\{\exp\left[-\frac{1}{4x_k^*}\left(\frac{z}{h}+\tau_k^* x_k^*-1\right)^2\right]+\alpha_k\exp\left[-\frac{1}{4x_k^*}\left(\frac{z}{h}+\tau_k^* x_k^*+1\right)^2\right]\right\}, \tag{9.43}$$

a modified Sutton equation including fall-out. In eqn. (9.43), D_{kx}, D_{ky}, D_{kz} are components of diffusivity of species k with coordinates x in the direction of flow and x vertically upward from ground level, while y is normal to both and symmetric with respect to plane x–z containing the center plane of the plume with x measured from the equivalent source.[130] Other notations are: h, the plume height given by w, u, Reynolds number at the stack exit, Grashof number and Richardson number of the atmosphere; α_k, the ground reflection coefficient; $\alpha_k = 1$ for non-absorbing ground, $\alpha_k = 0$ for a completely absorbing ground; τ_k^*, x_k^* are given by

$$\left.\begin{aligned}\tau_k^* &= \tau_k gh/D_{kz},\\ x_k^* &= D_{kz}x/uh^2,\end{aligned}\right\} \tag{9.44}$$

for constant D_k, g is the gravitational acceleration (g again can be replaced by $g+E_z\ (q/m)$ as in section 7.4), τ_k is the relaxation time for momentum transfer between the principal fluid of the plume and species k; $\tau_k g$ is its terminal velocity. For diffusivities given in the Sutton diffusion equation, or

$$\left.\begin{aligned}D_{ky} &= \frac{2-n}{4}\,C_{ky}^2 ux^{(1-n)},\\ D_{kz} &= \frac{2-n}{4}\,C_{kz}^2 ux^{(1-n)},\end{aligned}\right\} \tag{9.45}$$

where C_k's and n are constants,[130] eqn. (9.43) is still applicable for

$$x_k^* = \tfrac{1}{4}C_{kz}^2 x^{(2-n)}/h^2 \tag{9.46}$$

provided

$$\tau_k^* = 4\tau_k gh/(2-n)C_{kz}^2 ux^{(1-n)} \tag{9.47}$$

is a constant. Under this specification, eqn. (9.43) is applicable to cases of constant D_k or cases where D_k is a function of x.

The ground level flux J_{kG} of species k is given by diffusion as well as fall-out. Derivation based on eqn. (9.43) gives, in dimensionless form,

$$\begin{aligned} J_{kG} &\equiv J^*_{kGF}+\alpha_k J^*_{kGR} \\ &= \frac{1}{2}\left(\frac{\tau^*_k x^*_k+1}{x^{*2}_k}\right)\exp\left[-\frac{(\tau^*_k x^*_k-1)^2}{4x^*_k}\right] \\ &\quad+\frac{\alpha_k}{2}\left(\frac{\tau^*_k x^*_k-1}{x^{*2}_k}\right)\exp\left[-\frac{(\tau^*_k x^*_k+1)^2}{4x^*_k}\right], \end{aligned} \tag{9.48}$$

where J^*_{kG} is defined by, for constant D_k,

$$J^*_{kG}=\frac{J_{kG}h^2}{\pi r_1^2\rho_{k1}w_1}\left(\frac{4\pi uh}{D_{kz}}\right)\left(\frac{D_{ky}}{D_{kz}}\right)^{1/2}\exp\left[\frac{1}{4x^*_k}\frac{y^2}{h^2}\frac{D_{kz}}{D_{ky}}\right]. \tag{9.49}$$

For cases with Sutton diffusivity,

$$J^*_{kG}=\frac{J_{kG}}{r_1^2\rho_{k1}w_1}C_{ky}C_{kz}x^{2-n}\exp\left[\frac{1}{4x^*_h}\frac{y^2}{h^2}\frac{C_z^2}{C_y^2}\right], \tag{9.50}$$

where J^*_{kGF} is the dimensionless fall-out flux, and J^*_{kGR} is the dimensionless reflected flux.

Noting that at large τ^*_k, the reflected flux is negligible. The maximum J^*_{kGF} occurs at $x^*_k = 1/8$ at very small τ^*_k. As gravity effect increases or at large τ^*_k, the maximum J^*_{kGF} shifts towards lower values of x^*_k. It is interesting to note that J^*_{kGR} may have either positive or negative values because of the simultaneous effect of fall-out and diffusion, at $\tau^*_k = 0$, $\alpha^*_k = 1$, $J^*_{kG} = 0$ as in the case of the original Sutton equation, that is, no ground level flux.

The diffusion and fall-out from a jet engine at take-off can be designated by coordinate x along the ground level in the vertical plane of take-off at angle of ascent θ with ground at speed u. Here the problem becomes one of moving source. The correspondence to the case of the stack plume is seen in that now:

$$\left.\begin{aligned} x^*_k &= \frac{D'_{kz}(ut-x\cos\theta)}{ux^2\sin^2\theta}, \\ \tau^*_k &= \frac{\tau_k g\cos\theta\, x\sin\theta}{D'_{kz}}, \end{aligned}\right\} \tag{9.51}$$

and the variation of deposition rate at a given x with time t is analogous to the change of position along x in the case of the stack D'_{kz} is the diffusivity in the direction normal to the direction of ascent. Here we again account for only conditions after the effect of the jet velocity w itself has been diffused.

An example of using the charts based on eqn. (9.48) is for the conditions: $u = 100$ m/sec, $x = 2000$ m, $\theta = 30°$, $D_k = 10$ m²/sec, $\tau_k = 10^{-2}$ sec, $g = 9.80$ m/sec²,

$$\tau^*=\frac{10^{-2}\times 9.8\times 0.866\times 2000\times 0.5}{10}=98,$$

$$x^*=\frac{10^2}{2000\times 10(0.5)^2}\left[\frac{100t}{2000}-0.866\right]=2\times 10^{-4}[t-17.32].$$

The effect of reflection is negligible for this case. It is readily shown that fall-out began to be felt at $x^* = 10^{-2}$ or at $t = 22.3$ sec after take-off, reaching a maximum at $x^* = 0.005$ or $t = 42.5$ sec and nearly end of fall-out at $t = 167.3$ sec after take-off. For a jet of total exhaust flow of 100 kg/sec with 2% of the contaminant under consideration, $\pi r_1{}^2 w_k = 2$

kg/sec the maximum fall-out rate at $x^* = 0.01$ reaches $J_{kG}^* = 10^4$ or $J_{kG} = 0.16$ kg/m^2 sec as a maximum.

However, this is not the case of particles with $\tau_k = 10^{-3}$ sec; when other parameters are similar, fall-out may still be felt at $t = 2 \times 10^4$ sec or nearly 6 hr after take-off. It is readily seen that within up to 10 miles of a jet port, even though the number of flights might be reduced to a negligible amount over, say, 3 hr of each day, the fall-out of species from jet exhausts actually continues over the whole 24 hr.

CHAPTER 10

MOTION IN A STRONG ELECTRIC FIELD

When dealing with motion of a cloud of particles of large charge to mass ratios (say, 0.1 coul/kg or higher) in an intense applied electric field such that corona discharge may exist, the assumption of fluid forces greater than electrostatic forces in Chapter 9 has to be abandoned. Collision of charged species (gas ions and particulate matter) in the mixture on the neutral gas molecules will lead to electric wind, and the density distribution of particles is influenced by electric wind, diffusion, and deposition. This situation applies to electrostatic precipitators for air pollution control, atmospheric sampling, and cloud physics. We shall illustrate with cases of precipitators, on which definitive electrohydrodynamic analyses are available. The basic concepts are applicable to situations such as in cloud physics, where prominent electrical effect is produced by hydrodynamic transport.

10.1. Electric wind

A conventional one-stage precipitator (Fig. 24) consists of a metal duct through which the gas flows with one or more fine wires placed along the center of the duct. The wires are charged electrostatically to a high potential with respect to the duct walls and are used to charge the small particles in the flowing gas. This potential, being not far below the spark-over potential between the wire and the duct, is sufficiently high to produce in the region near the wire a high degree of ionization commonly referred to as "corona". The particles in the gas collect charges of the same polarity as the wire. These charged particles are then attracted towards the walls of the duct by the electric field. They migrate through the gas in the direction of the electric field away from the ionizer wire to the duct wall where they become attached and retained.

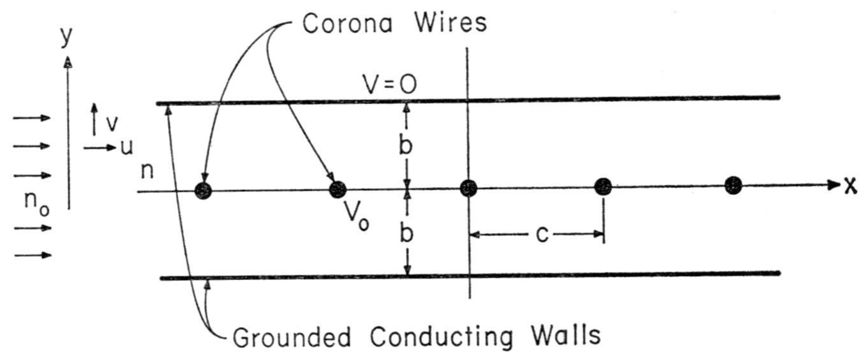

FIG. 24. Components of an electrostatic precipitator and coordinates.

The phenomenon known as the electric wind refers to the movement of gas induced by the repulsion of ions from the neighborhood of a high voltage discharge electrode. The effect is most commonly observed with atmospheric air, but it occurs also in other gases and insulating liquids. Ladenburg[123] investigated this effect in the electrostatic gas cleaning of a cylindrical configuration.

Robinson reviewed some of the methods of approach that have been used to treat this problem and reported recent measurements.[124] Also, by an approximate theory which relates electrodynamic quantities, he examined the role of electric wind in the charging electrode system.[125]

Since the population of gaseous ions far exceeds that of the dust particles, Ramadan and Soo[126] treat electric wind due to ions alone. For an infinitely long plate-wire combination, at steady state, the ion generation and recombination in each element is repeated. Computation based on such a two-dimensional element is therefore representative.

Ionization is produced by the corona field. The first Townsend ionization coefficient α in ion pairs/m is based on ion generation between two flat electrodes in a cylindrical column of radius R and length x.[127] At steady state, with ion mobility K and longitudinal electric field E, the number of ions generated in the column $\alpha n\,dx$ having a velocity KE is equal to those lost by drift at velocity v_d in the radial direction, or α is given by the relation α/P as a function of E/P for many gases; P is the pressure. The linear approximation $(\alpha/P) = \beta_1(E/P)$ is applicable over a wide range; $\beta_1 = 0.015$ ion pair/volt for air between E/P of 200 to 400 v/cm mmHg. For low extent of ionization in a corona where charges of one sign predominate, balance of ion generation and drift gives the charge density of ions:

$$\rho_i(q/m) \simeq \varepsilon_0(\beta_1/2\pi)E^2 \equiv \varepsilon_0 \beta E^2 \tag{10.1}$$

as an approximation. β can be treated as an empirical constant having a unit of v^{-1}. Further, β is expected to be smaller than $(\beta_1/2\pi)$ due to diffusion in a flow system, and, in general, ion density is dependent on the flow velocity.

For low extent of ionization, the density ρ of the principal fluid can be treated as incompressible although the ion density varies[77] and the overall continuity relation is given by

$$(\partial u/\partial x)+(\partial v/\partial y) = 0, \tag{10.2}$$

where u and v are components of secondary flow velocity in the x- and y-directions (Fig. 24) for principal slug flow of velocity U in the x-direction. The ion velocity is given by $v_i - v = KE$; $K = (q/m)F^{-1}$ is the mobility, and F is the inverse relaxation time of momentum transfer from ions to gas molecules. In non-dimensional form the momentum equations are

$$(1+u^*)(\partial u^*/\partial x^*)+v^*(\partial u^*/\partial y^*) = \rho_i^* E_x^*/(N_{Re})_e, \tag{10.3}$$

$$(1+u^*)(\partial v^*/\partial x^*)+v^*(\partial v^*/\partial y^*) = \rho_i^* E_y^*/(N_{Re})_e, \tag{10.4}$$

where, for cases $b = c$ in Fig. 24, $u^* = u/U$, $v^* = v/U$, $x^* = x/b$, $y^* = y/b$, $E^* = Eb/V_0$, $\rho_i^* = \rho_i/\rho_{i0}$, where ρ_{i0} is ρ_i at $E = V_0/b$. $(N_{Re})_e$ is the electric Reynolds number (inertia force/electrostatic force), $(N_{Re})_e = \rho U^2/\rho_{i0} V_0(q/m)$.[1]

The electric field E^* is given by the Poisson equation

$$\begin{aligned}(\partial^2 V^*/\partial x^{*2})+(\partial^2 V^*/\partial y^{*2}) &= -(\rho_i/\varepsilon_0)(b^2/V_0)(q/m)\\ &= -\beta V_0 E^{*2},\end{aligned} \tag{10.5}$$

which is reducible to the Laplacian form by the transformation

$$\varphi = (e^{\beta V_0 V^*}-1)/\beta V_0 \tag{10.6}$$

or

$$(\partial^2\varphi/\partial x^{*2})+(\partial^2\varphi/\partial y^{*2}) = 0 \tag{10.7}$$

and

$$V^* = (\beta V_0)^{-1} \ln(1+\beta V_0\varphi),$$
$$E_x^* = -(1+\beta V_0\varphi)^{-1}(\partial\varphi/\partial x^*).$$

The boundary conditions are: at wire of radius a, $V = V_0$, at $y = b$, $V = 0$. Solution to eqn. (10.7) was given by Cooperman for the case without space charge effect.[14] Equations (10.2) and (10.3) are now solved numerically with the boundary conditions:

$$u^* = 0, \text{ at } x = 0,\ y = 0 \text{ to } 1;\ x = 1,\ y = 0 \text{ to } 1.$$
$$v^* = 0, \text{ at } y = 0,\ x = 0 \text{ to } 1;\ y = 1,\ x = 0 \text{ to } 1.$$

Computations for various values of $(N_{Re})_e$ were made on an IBM 7094 computer, for both $\beta = 0$ (no space charge effect) and $\beta V_0 = 0.25$. Figure 25 shows the latter case for $(N_{Re})_e = 0.006$. A secondary flow was seen to be relative to the principal motion in the

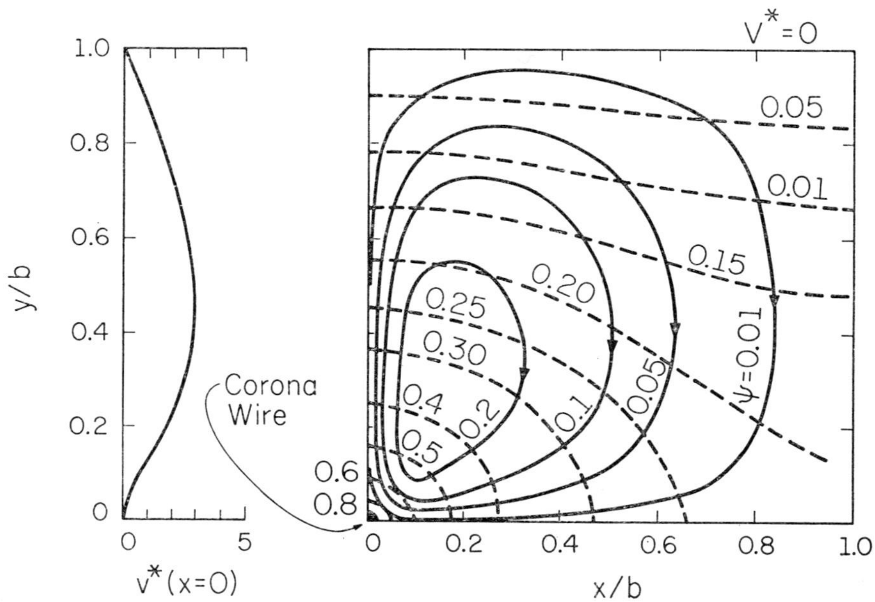

FIG. 25. Electrohydrodynamic secondary flow relative to principal motion for case $(N_{Re})_e = 0.006$, $b = c$ $\beta V_0 = 0.25$. Numerical computations made by O. E. Ramadan.

manner of narrow jet-like electric wind issued from the corona wire toward the plate. The space charge effect was shown to shift the peak of the jet velocity. In Fig. 25 the streamlines are given by stream functions ψ^*, with $\partial\psi^*/\partial y^* = u^*$, $-\partial\psi^*/\partial x^* = v^*$. The influence of $(N_{Re})_e$ is summarized in Fig. 26 for the v^*_{max} as in Fig. 25. It is seen that the jet from the electrode has a higher velocity at low electric Reynolds numbers. It is interesting to note that the electric wind in an electrostatic precipitator may have peak velocity several times that of the mean flow velocity, usually around 2 m/sec. Measurement using kerosene smoke tracer (7 μ mean diameter) gave results with range as marked on Fig. 26 for $V_0 = -25$, -30, and -40 kv.[128]

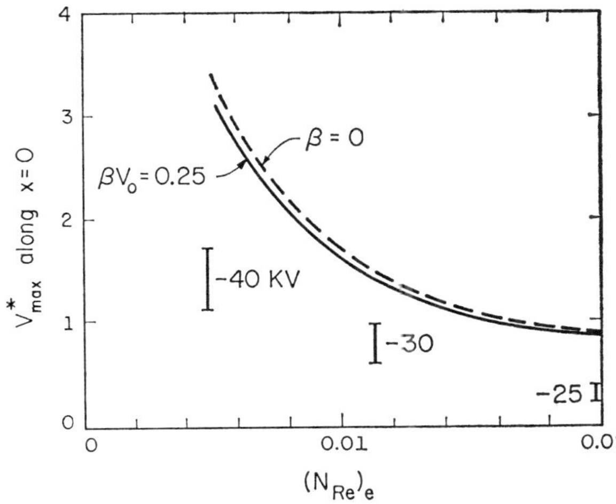

FIG. 26. Maximum electric wind velocity from corona wire to grounded plates and range of experimental results. Measurements made by O. E. Ramadan.

10.2. Density distribution in an electrostatic precipitator

The density distribution of dust particles in a precipitator (Fig. 24) is influenced by the interaction of electric field, diffusion, and deposition.[129] Following the basic relations in Chapter 6, Soo derived the basic relations and the perturbation method.

The total flux J of particles in a fluid is given by

$$\mathbf{J} = -D\nabla n + n(\mathbf{v}_n - \mathbf{v}), \tag{10.8}$$

where D is the diffusivity of particles, n is the number density, $\mathbf{v}_n$ its velocity of mass motion, and $\mathbf{v}$ is that of the fluid–particle mixture. For a dilute particulate suspension, $\mathbf{v}$ is nearly that of the fluid component. $\mathbf{v}_n$ is further given by[1]

$$d\mathbf{v}_n/dt = F(\mathbf{v} - \mathbf{v}_n) + (q/m)\mathbf{E}, \tag{10.9}$$

where t is the time coordinate, F is the inverse relaxation time of momentum transfer for particle fluid, (q/m) is the charge to mass ratio of a particle with q collected from the corona discharge, and E is the electric field. For small inertia force of particles compared to viscous force and electrostatic force, $d\mathbf{v}_n/dt = 0$ nearly, and the particle flux is given by

$$\mathbf{J} = -D\nabla n + (q/m)F^{-1}n\mathbf{E} \tag{10.10}$$

noting that $(q/m)F^{-1} = K$, the mobility of the charged particles.

The distribution of density of particles is given by

$$dn/dt = -\nabla \mathbf{J} \tag{10.11}$$

and the electric field is given by the Poisson equation

$$\nabla \mathbf{E} = \varepsilon_0^{-1} nq. \tag{10.12}$$

When applied to a two-dimensional element of an electrostatic precipitator as shown in Fig. 24, assuming constant q, m, and F for all the particles, eqns. (10.12) and (10.11) take the form for potential V

$$(\partial^2 V/\partial x^2) + (\partial^2 V/\partial y^2) = -\varepsilon_0^{-1} nq, \tag{10.13}$$

$$(u+KE_x)(\partial n/\partial x)+(v+KE_y)(\partial n/\partial y) = D(\partial^2 n/\partial y^2)+D(\partial^2 n/\partial x^2)-\varepsilon_0^{-1}Kqn^2, \quad (10.14)$$

where u, v are components of fluid velocity, and $E_x = -\partial V/\partial x$, $E_y = -\partial V/\partial y$.

The boundary conditions are:

$$\left.\begin{aligned} &y = 0, \quad \partial n/\partial y = 0, \quad V = V_0 \text{ at each wire} \\ \text{and} \quad &D\nabla n - K\mathbf{E}n = 0; \quad y = b, \quad V = 0, \end{aligned}\right\} \quad (10.15)$$

and at a given x over the whole section; the net loss of particles is equal to the displacement by the electric field and deposition at the wall,

$$\int_0 (n_0-n)u\, dy = \eta \int_0^x Kn_w E_{yw}\, dx, \quad (10.16)$$

when n_0 is the uniform particle density at the inlet; subscript w denotes the wall ($y = b$), and η is the attachment coefficient ($\eta = 1$ for complete attachment of particles reaching the wall).

Perturbation method: Eqns. (10.13) to (10.16) are non-dimensionalized by introducing $V^* = V/V_0$, $n^* = n/n_0$, $x^* = x/b$, $y^* = y/b$, $u^* = u/u_0$, $v^* = v/v_0$, and we further introduce the space-charge parameter (potential due to space charge/applied potential)

$$N_{SV} = n_0 q b^2/\varepsilon_0 V_0 \equiv A;$$

the diffusion parameter

$$N_D = bu_0/D \equiv B;$$

the drift-diffusion parameter

$$N_{DV} = (q/m)F^{-1}V_0/D = KV_0/D \equiv C;$$

and the space charge-diffusion parameter

$$H = AC.$$

Equation (10.13) was solved successively with the following perturbations:

$$V^* = V^{(0)}+AV^{(1)}+A^2V^{(2)}+\ldots, \quad (10.17)$$

$$n^* = n^{(0)}+Hn^{(1)}+H^2n^{(2)}+\ldots. \quad (10.18)$$

Computations were carried out on the above equations, under the assumption of small electric wind effect and one-seventh velocity law of the fluid phase and complete sticking ($\eta = 1$) of particles displaced to the wall. Results were comparable to results of measurement made with a Sinclair–Phoenix photometer on oil smoke introduced into an experimental precipitator section,[128] and checks the trend given by Robinson.[124]

Slow convergence of eqn. (10.18) at large H calls for direct computation of eqns. (10.13) to (10.16) simultaneous with the equations of motion.

10.3. Flow into a channel with corona charging

The first stage of a two-stage electrostatic precipitator element usually consisted of a corona wire and a pair of flat plates as shown in Fig. 27,[131] similar to the one used in section 9.2. Figure 27 shows also the equipotential line of the system of duct walls, plates, and the corona wire. The notations are similar to those in section 9.2. In this case the charge to mass ratio was at the level of 0.1 coul/kg for the results in Fig. 28, giving the mass flow and density distribution of particles for various values of x. It is noted that in the entrance region, the profiles are significantly different from the case without the corona field.[120] The effects include electric wind, large field force, and the wake of the wire.

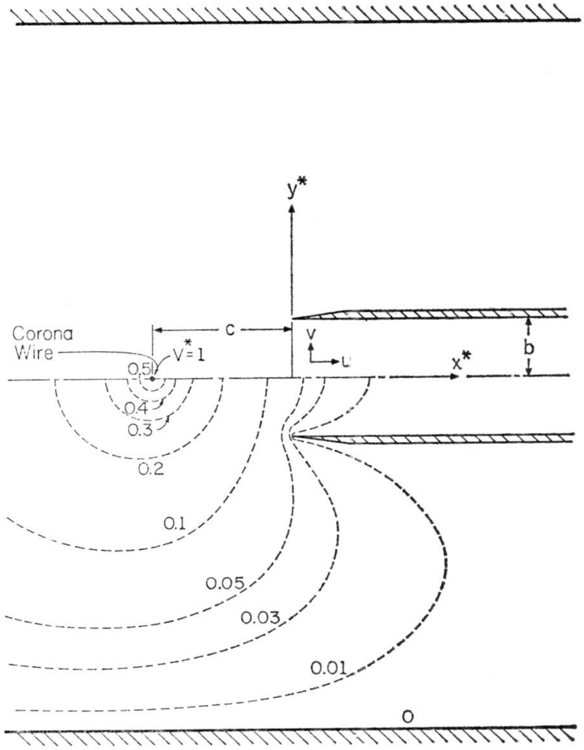

FIG. 27. Flow into a channel with corona charging ($x^* = x/b$, $y^* = y/b$). Equipotential lines computed by J. M. Hughes.

10.4. Electrification of clouds

With the accumulated knowledge on electrohydrodynamics of suspensions, we are still a long way from rational prediction of thunderstorms.[3,133] Here we are again dealing with situations where the electrical effect is prominent although the source is the hydrodynamic effect. Observations and measurements are difficult because of the elusiveness of thunderstorms and lightnings. Over the past 50 years theories have been limited to explanations of observed phenomena. These results are summarized here to indicate the orders of magnitude of events. Further understanding may aid future prediction and modification of weather.

Field change

In the fine-weather condition, the earth's field is nearly +120 v/m. Measurements made with a capillary electrometer[3] show that a distant lightning might produce a sudden large positive field change followed by a slow exponential recovery. Treating a thundercloud as a bipolar electrostatic generator with positive and negative centers of charges, charge $+q$ at height h_1 above $-q$ at h_2. At an observation point at a large distance L, it was shown

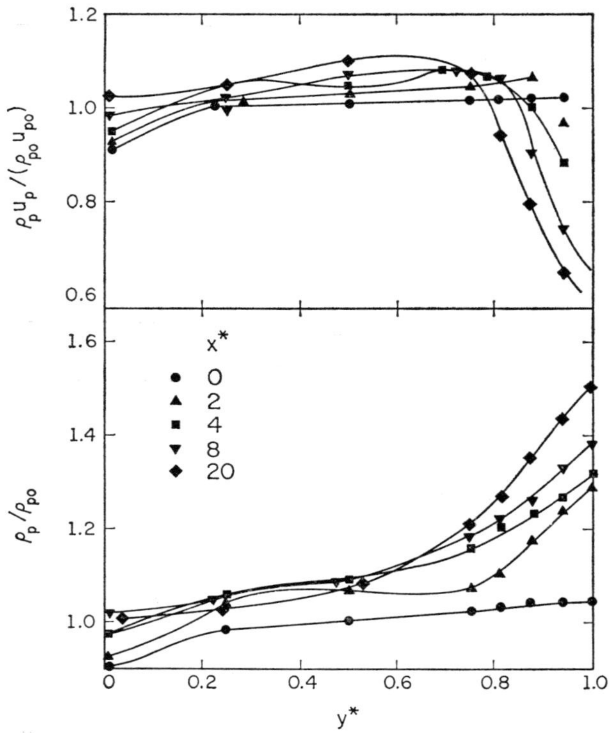

FIG. 28. Mass flow and density distribution of particles in the channel. $V_c = 35$ kv, $b = 1$ in., $c = 2\frac{1}{4}$ in., $\dot{m}_{p0} = 0.0286$, $\rho_{p0} = 2.14\times 10^{-3}$ lb m/ft^3. Measurements made by J. M. Hughes.

that for both cloud-to-ground discharge from h_2 and internal flashes the field-change is approximated by

$$\delta E \sim \delta M/4\pi\varepsilon_0 L^3, \tag{10.19}$$

where δM is the change in electric moment when a discharge δQ occurred. The polarity of the thundercloud was determined by varying L such that for internal flashes δE given by the following relation changes sign

$$\delta E = \frac{(\delta Q)}{2\pi\varepsilon_0}\left[\frac{h_2}{(h_2^2+L^2)^{3/2}} - \frac{h_1}{(h_1^2+L^2)^{3/2}}\right]. \tag{10.20}$$

$\delta E > 0$ at small L and <0 at large L was confirmed by most observations and hence $+q$ above $-q$ in most cases. Cloud-to-ground discharge was positive at all distances. The proportions of the frequency of occurrence of these two kinds of flashes vary with location, however. For L of 20 to 200 km, δM was nearly 100 C km and h_2 varied from 1.5 to 3.5 km, and $h_1 - h_2$ was 3–8 km. 10–30 C were neutralized by each internal discharge.

Electric fields in the interval between successive strokes of a flash were measured. A cloud-to-ground discharge is initiated by a streamer in steps of 10–200 m long and 30–100 μsec in each step. Photographic records gave such leader stroke at 4×10^5 m/sec and field change measurements gave 6×10^4 m/sec. When the leader stroke reaches the ground trailed by a channel of ionized gas (2–5 mm diameter), return stroke up through the channel occurred with great luminosity and speed (3×10^7 m/sec). Current was 2×10^4 amp on the average, while 2.2×10^5 amp were observed. A dart leader stroke may follow after 10^{-2} sec with streamer at 2×10^6 m/sec. Successive strokes may last for $\frac{1}{4}$ sec and up to 2 sec. Total charge neutralized might reach 164 C, with average of 30 C of which 20 C comes to earth. It was found[133] that 4–10% of discharges to ground contain positive charges, especially toward the later stages of thunderstorms, up to 100 C and up to 1.8×10^5 amp. The average change of electric moment in various section of lightning flash ranges from -25 C km to -35 C km.

Distant flashes

The dynamics of electrified clouds is seen in the recovery of the field after a discharge. Intense positive field may occur beneath thunderstorms accompanied by heavy rainfall, although such pre-discharge fields are predominantly negative. For a storm of 15–20 km distant, there is an enhancement of normal positive field of say 10^4 v/m, whereas field change produced by a lightning flash may be greater than 5×10^4 v/m. Recovery follows each discharge, with a relaxation time of about 7 sec for such a storm. For the case where the cloud is producing internal flashes, each destroys a bipolar charge distribution, and the electric moment is regenerated thereafter. Before the discharge, a field of -10^4 v/m may exist and therefore a positive space charge between the cloud base and ground. A positive field change produces an intense positive field at the ground because of positive space charge. The latter will be driven into the ground by the field or become neutralized by a rush of negative ions from available point discharges. This, together with the regeneration of the electric moment in the cloud, will cause a rapid initial decay of the field. Where there is no local point discharges or space charge the exponential recovery of field represents the regeneration of electric movement in the cloud. To generate a large electric movement, particles of different sizes and charge must be separated over a large vertical distance at velocity of fall of the large particles (section 7.3). The initial rate of regeneration would be nearly 100 C km in 7 sec, giving a vertical separation of $(7v)$ m (v is velocity in m/sec); the charges separated after the discharge is $(800/v)$ C. For a terminal velocity of 8 m/sec (section 4.5) 1000 C would be involved. This is visualized as a large neutral volume of cloud with 1000 C of mixed charges which as separation proceeds, gives a cloud with top at $+20$ C and bottom at -20 C until an internal field sufficiently intense to cause a flash is attained.

Over a recovery time of 20 sec, the rate of separation is the rate of fall under the influence of the internal field, or

$$v' = v_0 - \alpha M = \alpha(M_0 - M), \tag{10.21}$$

when v_0 is the velocity of free-fall without an electric field, M is the electric moment, and α, M_0 are constants. In the meantime loss occurs by convection, point discharges, etc., such that the rate of generation of moment is

$$\frac{dM}{dt} = Qv' - \beta M = \alpha Q(M_0 - M) - \beta M, \tag{10.22}$$

where Q is the total negative charge on the larger precipitation particles in the whole volume and β is the inverse relaxation time for discharges. With boundary condition of $M = 0$ at $t = 0$,

$$M = \frac{\alpha M_0 Q}{\alpha Q + \beta}\{1 - \exp[-(\alpha Q + \beta)t]\}, \tag{10.23}$$

a trend which was generally observed. $(\alpha Q+\beta)$ is thus the inverse relaxation time of the regeneration process. Note that $v_0 = \alpha M_0 = g/F$, while $\alpha M = KE = (q/m)F^{-1}E$ or $\alpha \sim K/h^3$, where K is the mobility of the falling particle and h is the characteristic separation distance.

Near flash

For flashes closer than 3 km, the change in electric field follows a different form from that for more distant flashes. In a region of widespread point-discharge currents, the current density is given by

$$J = neKE, \tag{10.24}$$

where n is the number density of ions of charge e. The Poisson equation takes the form

$$\frac{dE}{dh} = \frac{ne}{\varepsilon_0}, \tag{10.25}$$

where h is the height. Substitution of eqn. (10.25) into eqn. (10.24) followed by integration gives

$$E^2 - E_0^2 = 2Jh/\varepsilon_0 K, \tag{10.26}$$

where E_0 is the field at the ground. Further the potential is given by

$$V = \left(\frac{8J}{9\varepsilon_0 K}\right)^{1/2} [(h+c)^{3/2} - c^{3/2}] \tag{10.27}$$

with

$$c = \varepsilon_0 E_0^2 K/2J.$$

Note that both E_0 and c are small. High electric field immediately after the flash was caused by rapid dispersal of the space charge. Owing to a near balance between the effects of negative charge in the base of the cloud and the positive space charge below, the electric field before the flash is small. The discharge now removes the negative charges and leaves a field E due to space charge. This E is related to the surface charge density σ on the earth:

$$\sigma = \varepsilon_0 E. \tag{10.28}$$

Since

$$J = \frac{\varepsilon_0 K}{2h} E^2 \tag{10.29}$$

and

$$J = -\frac{d\sigma}{dt} = -\varepsilon_0 \frac{dE}{dt}, \tag{10.30}$$

we get

$$\frac{dE}{dt} = -\frac{J}{\varepsilon_0} = -\frac{K}{2h} E^2. \tag{10.31}$$

Thus for close discharges the recovery should proceed at a rate proportional to E^2.

Direct observations of thunderstorms by Kuettner[133] with measurement by an alti-electrograph show simultaneous occurrence of snow, hail or graupel, heavy shower, and moderate rain, and that the charge distribution was strongly influenced by the temperature distribution.

REFERENCES

1. Soo, S. L., *Fluid Dynamics of Multiphase Systems*, Blaisdell, Waltham, Massachusetts (1967).
2. Rudger, W. A. D., *Phil. Mag.* **25**, 481 (1913).
3. Mason, B. J., *The Physics of Clouds*, Oxford University Press, pp. 20, 22, 108, 110, 420; chapter 10 (1957).
4. Montgomery, D. J., *Solid State Physics*, **9**, 139 (1959).
5. Cheng, L. and Soo, S. L., *J. Appl. Phys.* **41**(2), 585–591 (1970).
6. Peterson, J. W., *J. Appl. Phys.* **25** (4), 501–504 (1954).
7. Penney, G. W. and Klingler, E. H., *Comm. Elec. AIEE*, July 1962, pp. 200–204.
8. Gilbert, L. and Hobbs, G. R., *Nature*, **195**, 275–276 (1962).
9. Thrane, E., *Electron Density Distribution in Ionosphere and Exosphere*, Wiley, New York (1964).
10. Coroniti, S. C., *Problems of Atmospheric and Space Electricity*, Elsevier, New York, chapters 1 and 2, pp. 178, 410, 485 (1965).
11. Friedman, H., *Physics of Upper Atmosphere* (J. A. Ratcliff, ed.), Academic Press, New York, pp. 17, 133 (1960).
12. Takahashi, T., *J. Atm. Sci.* **23**, 74 (1966).
13. Pierce, E. T. and Whitson, A. L., *J. Atm. Sci.* **22**, 314 (1965).
14. White, H. J., *Industrial Electrostatic Precipitator*, Addison-Wesley, Reading, Massachusetts, pp. 97, 135 (1963).
15. Cobine, J. D., *Gaseous Conductors*, Dover, New York (1958).
16. Soo, S. L., Wu, C. and Dimick, R. C., Recombination coefficients and transport properties of an ionized suspension, *1968 Intersociety Energy Conversion Engineering Conference Record, IEEE*, pp. 915–925 (1968).
17. Hughes, A. L. and DuBridge, L. A., *Photoelectric Phenomena*, McGraw-Hill, New York, p. 381 (1940).
18. Fleagle, R. G. and Businger, J. A., *Atmospheric Physics*, Academic Press, New York, p. 170 (1963).
19. Schultz, R. D. and Wiech, R. E., Jr., Advanced propulsion techniques (S. S. Penner, ed.), *AGARD Combustion and Propulsion Panel*, Pergamon, New York, p. 98 (1961).
20. Henriquez, P. C., *Rev. Trav. Chim.* **54**, 574 (1935).
21. Arendt, P. and Kallmann, H., *Z. Phys.* **35**, 421 (1926).
22. White, H. J., *AIEE Trans.* **70**, 1189 (1951).
23. Penney, G. W. and Lynch, R. D., *AIEE Trans.* **76**, 294 (1957).
24. Murphy, A. T., Adler, F. T. and Penney, G. W., *AIEE Trans.* **78**, 318 (1959).
25. Jost, W., *Diffusion in Solids, Liquids and Gases*, Academic Press, New York, p. 46 (1952).
26. Murphy, A. T., Charging of fine particles by random motion of ions in an electric field, Ph.D. thesis, Carnegie Institute of Technology, Pittsburgh, Pennsylvania (1956).
27. Liu, B. Y. H., Whitby, K. T. and Yu, H. H. S., *J. Colloid Interface Sci.* **23**, 367–378 (1967).
28. Liu, B. Y. H., Whitby, K. T. and Yu, H. H. S., *J. Appl. Phys.* **38** (4), 1592–1597 (1967).
29. Rosen, G., *Phys. Fluids*, **5**, 737 (1962).
30. Dimick, R. C. and Soo, S. L., *Phys. Fluids*, **7**, 1638 (1964).
31. Green, H. L. and Lane, W. R., *Particle Clouds, Dusts, Smokes and Mists*, E. and F. N. Spon, London, p. 82, chapter 11 (1957).
32. Allis, W. P. and Rose, D. J., *Phys. Rev.* **93** (1), 83–93 (1954).
33. Gunn, R., *J. Geophys. Res.* **54**, 57 (1949).
34. Whitby, K. T. and Liu, B. Y. H., *Aerosol Science* (C. N. Davis, ed.), Academic Press, London, chapter 3 (1967).
35. Ergun, D. and Orning, A. A., *Ind. Eng. Chem.* **41**, 1179 (1949).
36. Schlichting, H., *Boundary Layer Theory*, McGraw-Hill, New York (1960).
37. Kraemer, H. R. and Johnstone, H. F., *Ind. Eng. Chem.* **47**, 2426 (1955).
38. Liu, V. C., Pang, S. C. and Jew, H., *Phys. Fluids*, **8**, 788 (1965).
39. Patterson, G. N., *AIAA Jl.* **3**, 577 (1965).
40. Saffman, P. G., *J. Fluid Mech.* **22** (22), 385 (1965).
41. Tchen, C. M., Dissertation, Delft, Martinus Nijhoff, The Hague (1947).
42. Smythe, W. R., *Phys. Fluids*, **4**, 756 (1965).
43. Elrick II, R. M. and Emrich, R. J., *Phys. Fluids*, **9** (1), 28–32 (1966).

44. Fuchs, N., *The Mechanics of Aerosols*, Macmillan, New York, pp. 56, 161 (1964).
45. Happel, J. and Brenner, H., *Low Reynolds Number Hydrodynamics*, Prentice-Hall, Englewood Cliffs, New Jersey (1965).
46. Marchildon, E. K., Clamen, A. and Gauvin, W. H., *Phys. Fluids*, **7**, 2018; *Can. J. Chem. Eng.* **42** 178 (1964).
47. Soo, S. L., Think two-phase, *Advances in Solid-Liquid Flow in Pipes* (I. Zandi, ed.), Pergamon Press pp. 35–38 (1971).
48. Taylor, T. D., *Phys. Fluids*, **6**, 978 (1963).
49. Springer, G. S. and Tsai, S. W., *Phys. Fluids*, **8**, 1561 (RN), (1965).
50. Kassoy, D. R., Adamson, T. C. Jr., and Messiter, A. F., *Phys. Fluids*, **9**, 671 (1966).
51. Garner, F. H. and Hammerton, D., *Chem. Eng. Sci.* **3**, 1 (1954).
52. Levich, V. G., *Physicochemical Hydrodynamics*, Prentice-Hall, Englewood Cliffs, New Jersey, p. 209 (1962).
53. Chao, B. T., *Phys. Fluids*, **5**, 69 (1962).
54. Ingebo, R. D., NACA TN 3762 (1956).
55. Winnikow, S. and Chao, B. T., *Phys. Fluids*, **3** (1), 50–61 (1966).
56. Conkie, W. R. and Savic, P., Nat. Res. Council (Canada) Rept. No. MT-23 (1953).
57. O'Brien, V., *Phys. Fluids*, **6**, 1356 (1963).
58. Strutt, J. W. (Lord Rayleigh), *Theory of Sound*, Dover, New York, pp. 370, 374 (1929).
59. Ailam, G. and Gallily, I., *Phys. Fluids*, **5**, 575 (1962).
60. Mellor, M., Light scattering and particle aggregation in snowstorms, Army Cold Regions Res. and Eng. Lab., Hanover, N.H. (Feb. 1963) (CRREL-RP-193, AD-633539).
61. Johnson, W. B., *J. Air Pollution Control Assoc.* **19** (3), 176–180 (1969).
62. Blitz, J. *Fundamentals of Ultrasonics*, Butterworths, London (1963).
63. Bernard, B. J. and Binnie, A. M., *J. Fluid Mech.* **9**, 35 (1963).
64. Peskin, R. L. and Wallace, J. P., Some statistical turbulence properties of the fluid phase in a gas solid suspension, Rutgers Univ. Tech. Rept. No. 112-ME-F, Contract No. AT(30–1)2930, NYO2930-9 U.S. AEC (Sept. 1966).
65. Hinze, J. O., *Turbulence*, McGraw-Hill, New York, pp. 185, 533 (1959).
66. Jastrow, R. and Pearse, C. A., *J. Geophys. Res.* **62**, 413 (1957).
67. Wyatt, P. J., *J. Geophys. Res.* **65** (6), 1673 (1960).
68. Chopra, K. P., *Rev. Mod. Phys.* **33**, 153 (1961).
69. Davis, A. H. and Harris, I., NASA TN D-704 (1961).
70. Licht, A. L., *J. Geophys. Res.* **65**, 3493 (1960).
71. Pitts, W. C. and Knechtel, E. D., NASA TN D-2619 (1965); *AIAA Jl.* **2**, 1148 (1964).
72. Hoult, D. P., *Phys. Fluids*, **8**, 1394 (1965).
73. Soo. S. L., *Advanced Heat Transfer* (B. T. Chao, ed.), University of Illinois Press, pp. 415–437 (1969).
74. Cheng, L., Tung, S. K. and Soo, S. L., *Trans. ASME, J. Eng. Power*, **42A** (2), 135–149 (1970).
75. Daniel, J. H. and Brackett, F. S., *J. Appl. Phys.* **22**, 542 (1951).
76. Min, K., Chao, B. T., and Wyman, M. E., *Rev. Sci. Inst.* **34** (5), 529–531 (May 1963).
77. Soo, S. L., *ZAMP*, **19**, 545–563 (1968).
78. Fletcher, N. H., *The Physics of Rain Clouds*, Cambridge University Press (1962).
79. Fukuta, N., *J. Atm. Sci.* **23**, 191 (1966).
80. Engelmann, R. J., *J. Atm. Sci.* **22**, 719 (1965).
81. Philip, J. R., *J. Atm. Sci.* **22**, 196 (1965).
82. Hallett, J., *J. Atm. Sci.* **22**, 64 (1965).
83. Latham, J. and Stow, C. D., *J. Atm. Sci.* **23**, 245 (1966).
84. Miller, A. H., Shelden, C. E. and Atkinson, W. R., *Phys. Fluids*, **8**, 1921 (1965).
85. Battan, L. J. and Thiess, J. B., *J. Atm. Sci.* **23**, 78 (1966).
86. Fujiwara, M., *J. Atm. Sci.* **22**, 585 (1965).
87. Lane, W. R., *Ind. Eng. Chem.* **43**, 1312 (1951).
88. Clark, B. J., NASA TN D-2424 (1964).
89. Hueter, T. F. and Holt, R. H., *Sonics*, Wiley, New York, p. 418 (1955).
90. Koenig, L. R., *J. Atm. Sci.* **22**, 585 (1965).
91. Twomey, S., *J. Atm. Sci.* **23**, 405 (1966).
92. Gunn, R., *J. Atm. Sci.* **21**, 168 (1964).
93. Soo, S. L., *Ind. Eng. Chem. Fundamentals*, **4**, 426 (1965).
94. Soo, S. L., *Proc. 17th Congress of Int. Astro. Fed., Madrid, Spain*, **3**, 3–11 (1966).
95. Soo, S. L., *Proc. Symposium on Single and Multi-component Flow Processes* (R. L. Peskin and C. F. Chen, eds.), Eng. Res. Pub. No. 45, Rutgers, New Brunswick, N.J. 1 (1965).
96. Stuetzer, O. M., *Phys. Fluids*, **5**, 534 (1962).
97. Lockwood, D. L. and Hamza, V., NASA TN D-1461 (1962).
98. Langmuir, I. and Blodgett, K., *Phys. Rev.*, Ser. 2, **24**, 49 (1924).

99. Lipscomb, W. N., Rubin, T. R. and Sturdivan, T. J. H., *J. Appl. Phys.* **18**, 72 (1947).
100. Owen, P. R., Proc. BCURA Conf. at Leatherhead, Surrey, England, *Aerodynamic Capture of Particles* (E. G. Richardson, ed.), Macmillan, New York (1960).
101. Shishkin, N. S., *Problems of Atmospheric and Space Electricity* (S. C. Coroniti, ed.), Elsevier, New York, p. 268 (1965).
102. Battan, L. J., *J. Atm. Sci.* **22**, 79 (1965).
103. Latham, J. and Miller, A. H., *J. Atm. Sci.* **22**, 505 (1965).
104. Moore, L. B., Vonnegut, B., Vrablik, E. A. and McCaig, D. A., *J. Atm. Sci.* **21**, 646 (1964).
105. Makhotkin, L. G., *Izv. Atm. Oceanic Phys.* **1**, 230 (p. 135 in English translation) (1965).
106. Pasquill, F., *Atmospheric Diffusion*, Van Nostrand, New York, p. 112 (1962).
107. Banister, J. R. and Davis, H. L., *Phys. Fluids*, **5** (2), 136 (1962).
108. Katz, E. J., *J. Atm. Sci.* **23**, 159 (1966).
109. Walker, E. R., *J. Appl. Meteorol.* **4**, 614 (1965).
110. Soo, S. L., Effect of simultaneous diffusion and fall-out from plumes of stacks and jet engines, Paper No. 70–81, *63rd Annual Meeting of the Air Pollution Control Association, St. Louis, Mo., June 14–19, 1970; J. Atmospheric Environment* (in press) 1971.
111. Soo, S. L. and Trezek, G. J., *Ind. Eng. Fundamentals*, **8** (1966).
112. Soo, S. L., *Appl. Sci. Res.* **21**, 68–84 (1969).
113. Soo, S. L., Stukel, J. J. and Hughes, J. M., *Environmental Sci. Tech., Ind. Eng. Chem.* **3** (4), 386–393 (1969).
114. Einstein, A., *Am. Phys. (Leipzig)*, **19**, 289 (1906).
115. Jeffery, G. B., *Proc. Roy. Soc. (London), AIIE*, 161 (1922).
116. Segré, G. and Silberberg, A., *Nature*, **189**, 209 (1961); *J. Fluid Mech.* **14**, 115, 136 (1962).
117. Soo, S. L., *Astronautica Acta* **11** (6), 422 (1965).
118. Liu, J. C. T., *Phys. Fluids*, **9**, 1716 (1966).
119. Miller, E., Thunderstorm lightning, *Proc. 13th Annual Air Force Science and Engineering Symposium AEDC, Sept. 1966*, paper No. 48.
120. Stukel, J. J. and Soo, S. L., *Particle Technology*, **2** (5), 278–289 (1969).
121. Soo, S. L., Electrohydrodynamic jets, *Proc. 11th Annual Midwestern Mechanics Conference, Iowa State University* **5**, 215–224 (1969).
122. Lee, Y. N. and Soo, S. L., A study of jets of electrically charged suspensions, *Environmental Sci. Tech., Ind. Eng. Chem.* **4**, 678–685 (1970).
123. Ladenburg, R., *Annalen der Physik, Leipzig, Germany* (5th ser.), **4**, 803–897 (1930).
124. Robinson, M., The effect of polarity on particle concentration profiles in electrostatic precipitators, *Jl. APCA*, **18** (10), 688 (1968).
125. Robinson, M., Movement of air in the electric wind of corona discharge, *AIEE Trans.* **80** (part 1), 143–150 (1961).
126. Ramadan, O. E. and Soo, S. L., *Physics Fluids*, **12** (9), 1943–1945 (1969).
127. Brown, S. C., *Basic Data of Plasma Physics*, Technology Press, John Wiley, p. 122 and fig. 4.37 (1968).
128. Ramadan, O. E., Stukel, J. J. and Soo, S. L., A study of the effects of electric wind in an electrostatic precipitator, *62nd Annual Meeting of the Air Pollution Control Association, June 23, 1969, New York, N.Y.* paper No. 69–1 (1969); also Ramadan, O. E., Ph.D. Thesis, Univ of Illinois, (1969).
129. Williams, J. C. and Jackson, R., The motion of solid particles in an electrostatic precipitator, *Proc. Symposium on Interaction Between Fluids and Particles* (London: Inst. Chem. Engrs.), 282–288 (1962).
130. Stern, A. C., *Air Pollution*, 2nd edn., Academic Press, vol. 1, pp. 240, 265, 266 (1968).
131. Hughes, J. M., Stukel, J. J. and Soo, S. L., A preliminary study of hydrodynamics of a two-stage electrostatic precipitator element, *61st Annual Meeting of the Air Pollution Control Association, St. Paul, Minnesota, June 26, 1968*, paper No. 68–103.
132. Eckert, E. R. G. and Drake, R. M., Jr., *Heat and Mass Transfer*, McGraw-Hill, New York, p. 192 (1959).
133. Chalmers, J. A., *Atmospheric Electricity*, 2nd edn., Pergamon Press, Oxford, ch. 12, 14 (1967).

AUTHOR INDEX

(Only the name of the first author of two or more collaborators is indexed. The numbers in italics relate to reference numbers, and the prefix letter refers to the review author, i.e. *F* is Fuchs, *S* is Soo. The numbers in ordinary type are page numbers.)

SUBJECT INDEX